CHEMISTRY RESEARCH AND APPLICATIONS

A CONCEPTUAL APPROACH TO THE TEACHING OF CHEMISTRY

CHEMISTRY RESEARCH AND APPLICATIONS

Additional books in this series can be found on Nova's website
under the Series tab.

Additional e-books in this series can be found on Nova's website
under the e-books tab.

EDUCATION IN A COMPETITIVE AND GLOBALIZING WORLD

Additional books in this series can be found on Nova's website
under the Series tab.

Additional e-books in this series can be found on Nova's website
under the e-books tab.

CHEMISTRY RESEARCH AND APPLICATIONS

A CONCEPTUAL APPROACH TO THE TEACHING OF CHEMISTRY

RAFFAELE P. BONOMO
GIOVANNI TABBÌ
AND
ALESSANDRO GIUFFRIDA

New York

Copyright © 2013 by Nova Science Publishers, Inc.

All rights reserved. No part of this book may be reproduced, stored in a retrieval system or transmitted in any form or by any means: electronic, electrostatic, magnetic, tape, mechanical photocopying, recording or otherwise without the written permission of the Publisher.

For permission to use material from this book please contact us:
Telephone 631-231-7269; Fax 631-231-8175
Web Site: http://www.novapublishers.com

NOTICE TO THE READER

The Publisher has taken reasonable care in the preparation of this book, but makes no expressed or implied warranty of any kind and assumes no responsibility for any errors or omissions. No liability is assumed for incidental or consequential damages in connection with or arising out of information contained in this book. The Publisher shall not be liable for any special, consequential, or exemplary damages resulting, in whole or in part, from the readers' use of, or reliance upon, this material. Any parts of this book based on government reports are so indicated and copyright is claimed for those parts to the extent applicable to compilations of such works.

Independent verification should be sought for any data, advice or recommendations contained in this book. In addition, no responsibility is assumed by the publisher for any injury and/or damage to persons or property arising from any methods, products, instructions, ideas or otherwise contained in this publication.

This publication is designed to provide accurate and authoritative information with regard to the subject matter covered herein. It is sold with the clear understanding that the Publisher is not engaged in rendering legal or any other professional services. If legal or any other expert assistance is required, the services of a competent person should be sought. FROM A DECLARATION OF PARTICIPANTS JOINTLY ADOPTED BY A COMMITTEE OF THE AMERICAN BAR ASSOCIATION AND A COMMITTEE OF PUBLISHERS.

Additional color graphics may be available in the e-book version of this book.

Library of Congress Cataloging-in-Publication Data

Bonomo, Raffaele Pietro.
 A conceptual approach to the teaching of chemistry / Raffaele Pietro Bonomo, Giovanni Tabbl, Alessandro Giuffrida (Dipartimento di Scienze Chimiche, Universit` degli Studi di Catania, Catania, Italy).
     pages cm
  Includes index.
  ISBN 978-1-62257-862-7 (soft cover)
 1. Chemistry, Physical and theoretical--Study and teaching. I. Tabbi, Giovanni. II. Giuffrida, Alessandro. III. Title.
  QD453.3.B66 2013
  541.071--dc23
                                           2012031715

Published by Nova Science Publishers, Inc. † New York

CONTENTS

Preface		vii
Summary		ix
Chapter 1	Purpose of Scientific Language and Theory	1
Chapter 2	What is Chemistry About?	11
Chapter 3	About the Concept of Valence	19
Chapter 4	About the Periodic Table	27
Chapter 5	About Atomic Theory	37
Chapter 6	About Thermodynamics	45
Chapter 7	Chemistry Education and Science Teaching	55
Chapter 8	Conclusion	63
Acknowledgments		71
Index		75

PREFACE

We present here a set of historical and conceptual considerations about General Chemistry Syllabus, which have been used as introductory lectures to the said course. These considerations can be addressed both to high school students having their first course in Chemistry (in this case in a simplified form, of course) and university students taking their first introductory course in Chemistry. There are six parts, the first deals with the peculiarities of scientific language and the importance of scientific theories as a progressive approximation to a true account of the physical world, the second with the singularity of a discipline like Chemistry in the effort to discover the rational organization of nature, the third with the valence concept and its development up to the concept of the chemical bond, the fourth with the comparison between Mendeleev's periodic table and the modern periodic table of elements, the fifth with the atomic theory (we devote particular attention to the understanding of reality based on the assumptions of a physical theory), the sixth with the origin and the development of thermodynamics, the conclusion of which are compared to the physical world as conceived and pictured in Newtonian dynamics.

These ideas attempt to provide answers to questions that have changed over time, because we often find students unaware of the cultural differences that arise from different approaches to the same scientific problem. Moreover, on the basis of our teaching experiences, we have arranged these ideas, which have resulted from an in-depth study of the cultural perspectives of many scientists who played a leading role in science, so as to make the salient points of Chemistry clear and so easily used. In other words, these considerations may be employed as outlines of lectures on specific issues, which are to be completed by adding pertinent explicative examples. Furthermore, some of them may serve as introduction to each topic, while others can be useful as concepts to be proposed later on or at the end of the topic, i.e. considering them as a conclusion.

Indeed, we regret that, when we were students, no one has addressed to ourselves these ideas relative to a historical and conceptual approach to the study of chemistry, not even our best lecturers, who were perhaps mainly concerned with exploring the details of their disciplines only.

To conclude our discussion of this subject, we will also give some ideas as to considering a cultural approach to the teaching of Chemistry, which we consider of particular importance and, finally, some remarks about the modern teaching of science and our proposal to improve it.

We are not people involved in educational studies, but scientists, in particular, inorganic chemists involved in the field of Inorganic and Analytical Biochemistry and don't want to compete with leading education scholars who do research in the field of Education and write a lot of very interesting things, sometimes not very concrete, about the way of teaching in their journals. But we love teaching and are lecturers too, and we could not be satisfied with proposing passively the themes contained in a General Chemistry program. Therefore, these lectures would not exist without all the hard work we have done in the past and the time spent on trying to find the best way to present information on these topics that we thought important for our students. It was a really satisfactory work and we would like to dedicate it to all teachers of scientific disciplines who not only get passionately involved in their work as teachers, but also are interested in promoting deep understanding of the key points of their discipline. The students' motivation and participation to the regular didactic activities of a scientific course need to be greatly stimulated, otherwise the communication of the discipline details might of course remain mere idle talk about a series of facts which are presented and do require no further explication.

Summary

This little book presents historical and-conceptual considerations about some of the topics of General Chemistry Syllabus used by the authors as introductory lectures to the said course. There are six parts, the first deals with the peculiarities of scientific language and the importance of scientific theories, the second with the singularity of a discipline like Chemistry, the third with the valence concept and its development up to the concept of the chemical bond, the fourth with a comparison between Mendeleev's periodic table and the modern periodic table of elements, the fifth with the atomic theory, the sixth with the origin and the development of thermodynamics the conclusion of which are compared to the picture of physical world as conceived by Newtonian dynamics.

On the basis of their teaching experiences, the authors have arranged these ideas, so as to make the salient points of scientific discipline like Chemistry clear and so easily used. In other words, their considerations may be employed as outlines of lectures on specific issues, which are to be completed by adding pertinent explicative examples and reading passages.

To conclude the discussion of this subject, the authors also give some ideas as to considering a cultural approach to the teaching of Chemistry, which they consider very important, and in general their proposal to improve the present way of teaching scientific disciplines.

Chapter 1

PURPOSE OF SCIENTIFIC LANGUAGE AND THEORY

In this chapter we will analyze the power and limitations of scientific language comparing it to the ordinary language. We will start asking the following question: what is the difference between ordinary and scientific language? In other words, is there any specific distinction between ordinary language and scientific language working towards a common objective, which is providing a knowledge of reality?

Certainly the two languages have cognitive claims towards the reality of our world and its communication. The ordinary language has a richness that allows to distinguish among different situations. For example, the word man can indicate different things, having a correspondence with other words like, teacher, pilot, student, worker, officer, trader and so on. With our words we are able to make statements that describe reality accurately and, therefore, allowing for its communication.

Every element or term of a language is produced by

a proto-language (a language that is ancestor of another language or group of languages) that is one of the ways to create a language by means of nouns, adjectives, verbs, adverbs, grammatical and syntactic rules. And this occurs in the same way as it does for the machine language (often called assembly language) of a computer, language that allows it to store information and use them during calculations [1] *(our translation).*

A description of reality must be communicable to others and this is obtained by connecting words into sentences through grammatical and syntactic rules that are based on our experience of the world. The role of syntax is to avoid the possible ambiguities and misinterpretations that may arise from assertions valid in a different context. In fact, if a phrase only used grammar constraints, it could be disconnected from any descriptive intent, as in the following examples: Paul is a prime number, that horse belongs to the left, oblique numbers are infinite. These are all phrases that are grammatically legal, but they do mean nothing. Speech allows one to build up signs, articulate thoughts and communicate emotions, all procedures that allow to express our feelings, record our experiences and propose our ideas. Therefore, a phrase can be considered legitimate when it corresponds to reality. In fact,

> *even within phrases that could be considered similar, we have a lexical, grammatical and syntactic richness that allows for different interpretations* [2] (our translation),

each of which gives rise to a substantially different meaning:

(a) Paul sold Peter a bicycle for 100 dollars
(b) the Paul's bicycle cost Peter 100 dollars
(c) the bicycle sold Peter made Paul earn 100 dollars.

All these sentences have, after all, the same meaning: a bicycle was sold by a person to another person, while the first deals with a common contractual situation, the second puts a great focus on the expense of the buyer and the third on the earnings of the seller.

Let us examine the following sentence: cat eats bird. It is a legitimate sentence because it corresponds to reality and we have experience of it ourselves. In the ordinary language the syntax is supported by tradition, but the event that a cat eats a bird could not happen for various reasons, including the fact that may be the cat has already eaten something and, thus, is not hungry, may be the bird is far away, may be cat and bird actually became friends. An event that is the opposite of what was expected to occur, happens if the bird is a bird of prey, like hawks and eagles. Figure 1 shows some of these possibilities.

Figure 1. From the left to the right, three of these possibilities are represented: cat likes to hunt bird, cat and bird have a friendly interaction, attack of hawks or eagles, respectively

Therefore, the assertions we usually make in the ordinary language contain a dose of ambiguity. They have a descriptive intent and allow for communication of reality, but do not allow us to make useful predictions. And this happens because

> *the same word (symbol) can correspond to several objects* [3, 4] (our translation).

For example when we talk about moon: are we considering the world of poetry, referring to its astronomical representation or claiming mental insanity? Even, when we talk about water are we thinking of washing clothes, swimming in the sea, tilling the soil, or considering the physico-chemical properties of a simple molecule like H_2O? The same reality in different contexts has several symbols: Mary is in the car, is Mary the driver or a passenger among others?

Now, since we pretend through scientific language to make predictions about our world, the above-mentioned ambiguity must be avoided or reduced as much as possible by building up a bijection between symbol and reality. In fact, going back to the last example (cat eats bird) and pretending to make a prediction, we are usually constrained to carry out measurements and express their results by means of numbers. Then we will begin to ask questions such as: how much does this cat weight? How much does this bird weight? How old is this cat? How hungry is this cat? Which is the distance between this cat and this bird? Answers to all these questions are numbers and, since the grammatical rules, which connect words of the ordinary language, become mathematical equations in the scientific language, we should find an equation that describes the behavior of the cat, in terms of probability, and can indeed be a reasonable answer to the question: is cat really going to eat bird? Actually, the solution of this equation would allow us to make a prediction on the behavior of the cat and on the end of the bird, or vice versa. Let us make an attempt indicating:

1. Cw, the weight of the cat;
2. Bw, the weight of the bird;
3. Ch, the hunger of the cat (let us consider a number that varies from 0 to 1, so as 0 corresponds to the situation in which the cat is not hungry and 1 to that of a cat having the maximum hunger);
4. CBd, the distance between the cat and the bird;
5. Ca, the age of the cat, expressed as years/1-(years -7)^2, taking into account the ability of the cat as a function of its age (within the first year the cat has not developed any precise hunting technique and about seven years possesses a mature technique).

A possible equation, in which P indicates the probability that the cat eats the bird, could take the following form:

$$P = \frac{Cw \times Ch \times Ca}{Bw \times CBd}$$

Anyway, some ambiguities are always present. In fact, if CBd becomes zero, the probability P has an infinite value, but we don't know if the cat eats the bird or vice versa. Moreover, if Bw becomes zero, even in this case the probability is infinite, but the cat eats nothing. Therefore, the hypothetical relationship has to be verified and tested by observation and, after verification, it would be possible to confidently state that it is the right equation. We ought to use the conditional, because its validity depends from the number of verifications.

This structure, symbols → equations → reality, works as the ordinary language does, the syntactic rules being the physical laws. Hence a mathematical model is a hypothesis, expressed in the formal language of mathematics, from which reality can be read. In this schematization, the mathematical model often introduces some concepts that are abstract and not really relevant in the experimental facts, as the variables Ch (cat hunger) and Ca (cat age).

We have emphasized that the verification process is very important, though in principle the prediction of an event is almost contained within the set of symbols used to describe it, so as it would be superfluous a comparison with real facts. It means that it is possible to visualize every reality in the space of symbols, thus, there must be a one-to-one correspondence, or bijection, between symbol and reality, but, unfortunately, this is not always true. It is almost evident that we can make several hypotheses and find other relationships that could adequately describe the behaviour of the cat in a similar way. Hence, there are surely different points of view that often take place in different mental universes not only in the case of the ordinary language but also within the physical sciences. This is why it is so important to compare hypothetical propositions with reality, because there is no physical theory that could exhaust reality, which remains richer than what we may build up in our space of symbols.

Furthermore, the events considered by a scientific theory are not all the facts, but only those that may be coherently treated within the formalism of the model. Many times in the history of science there have been choices which have been done on the basis of convenience (an *ad hoc* hypothesis) rather than following the logical development of the basic ideas in a theory.

This is the reason why now we will move to answer another question: what is a scientific theory?

In order to give a reasonable answer we will recall an episode (occurred more than twenty years ago) which involved the scientific community. We refer to the supposed Fleischmann and Pons experiment [5] of the late '80 years, in which these authors claimed that under some specific experimental conditions metals such as platinum, palladium or titanium (metals that were well known to chemists for the ability to absorb hydrogen on their surface with consequential catalytic activities) could favour a deuterium nuclear fusion reaction. We all asked ourselves two questions.

1. Could it be true? In other words is it possible that a reaction as the fusion nuclear reaction, for which the most technologically advanced countries in the world spend a lot of money (the experimental conditions to realize a fusion nuclear experiment involve the transformation of an inert gas into plasma) could occur at the surface of a particular metal and further at room temperature?
2. Which are the practical applications of such an experiment? Could we foresee the possibility of exploiting the "cold fusion" as a new energy source as our world suffers from a lack of renewable energy sources?

In the first question, which goes beyond the oversimplification introduced by mass-media (very often, what they assert is really misleading), there is a sincere wish to know. Therefore, it is necessary to repeat the experiment several times for the purpose of obtaining statistical reliability and avoiding it hasn't arisen by accident or mere chance. Furthermore, a great theoretical effort has to be made to understand what it is really occurring and which peculiar conditions can be reached within the crystal structure of those metals. We will start in the search for a physical theory, which must be a descriptive and predictive model of the phenomenon. It is necessary to test it by means of subsequent experiments and observations, the results of which will tend to confirm, improve or confute it. In other words, if science is, above all, knowledge, it must give an answer to the following question: if "cold fusion" happened (while some experiments tend to confirm it, others seem to confute it, but actually, the scientific community has taken the ultimate position that the "cold fusion" simply does not exist) which is the physical mechanism behind it?

Though this phenomenon was almost unknown, suddenly the question of its practical applications was asked. There was a rush to patent the "cold fusion" phenomenon without having verified not only its experimental details, but also without having looked for a theoretical basis. We have some reservations about stressing such a practical aspect of contemporary science and think we ought to put less emphasis on it, because it forgets that science is above all knowledge: it doesn't matter if it was true, but the important thing is that it works! The only reason is obviously to draw the power to manipulate reality. The technological research is a well different thing, because it is the search for practical applications to our world, once the scientific knowledge has been thoroughly assessed. Actually, the technological research is essential part of the adventure of scientific research, because science avails itself of advances in technology in order to deepen its knowledge.

The aim of science is the discovery of the effective structure of the physical world, hence all the entities (atoms, molecules, elementary particles, photons and so on) defined within its field (once experimentally verified) are to be considered integrating part of the reality of the physical world. It is almost true that the picture of this world could be approximate, but a physical concept pretends to describe the physical system to which it refers. Now it is important to understand what a scientific description is. Modern science depends by both the discovery and study of analogies and the construction of models. It wouldn't be possible to project an experiment without having an initial idea of the physical system under study. In order to make an example of these last assertions, let us consider the classical example of the Boyle's law (R. Boyle, 1627-91) on the behaviour of gases. A simple description of this experiment states that, when the temperature is kept constant, volume and pressure of a gas behave inversely proportional: $P \times V = constant$ at T = constant, actually, E. Mariotte (1630-84) did specify that temperature must be held constant so that the Boyle's law is often referred as Mariotte's law. In searching for the relationship between volume and pressure of a gas (in particular, air), Boyle has been guided by what he called atomistic philosophy. He thought that air was substantially constituted by tiny particles and, therefore, differently from liquids and solids, it could be easily compressed.

> *Compressing air simply would mean, from that point of view, the squeezing of empty space out of the volume, pushing the atoms closer together* [6].

However, the Boyle's law remains an experimental law that asserts a relationship between volume and pressure of a gas, but did not explain why gases would behave this way. Instead it is the kinetic theory of gases that explains the inverse proportionality between volume and pressure. When the gas is considered as composed by a large number of tiny particles that are far apart in an empty space and obeying the laws of classical mechanics, if the pressure of a gas arises from the bombardment of its particles against the inside walls of the container, it is possible to obtain (following the kinetic theory) that $P \times V = 1/3 \times N \times m \times u^2 = constant$ (N and m are the number and the mass of particles, respectively, and u^2 their average quadratic speed) if the temperature is kept constant. Therefore the Boyle's law has received a theoretical explanation by means of the kinetic theory of gases.

Now since the theory describes the system of N particles moving in an empty space, a scientist doesn't doubt the gas is effectively formed by tiny particles. In other words, the explicative model of theory becomes descriptive of what does really exist in the system. Therefore, what makes theory satisfactory is that it makes assertions about the existence of theoretical entities, in particular, the molecules of gases. Concluding we can state that the Boyle's law is only an empirical law, which doesn't explain why gases behave in a certain way. On the contrary, the kinetic theory of gases is highly explicative because it hypothesizes the existence of molecules that obey the general laws of motion. Hence, scientific theories give more satisfactory explanations, because they commit themselves in a description on how things behave in reality. On the other hand, if it were not so, it would be impossible to state whether or not a particular theory could be true or false [7].

REFERENCES

[1] Borghi, C. (1976). *Se volessimo vederci chiaro,* Milano, Italy: Jaca Book.

[2] Rigotti, E. (1987). *Implicazioni antropologiche di alcune concezioni correnti del rapporto tra lingua e testo.* Synesis: L'uomo e le sue espressioni, 3/4, 53-108.

[3] Arecchi, F. T. (1986). Conoscenza e realtà. *Synesis: Il modello matematico,* 2/3, 127-144.

[4] Arecchi, F. T., & Arecchi, I. (1990). *I Simboli e la Realtà.* Milano, Italy: Jaca Book.

[5] Fleischmann, M. and Pons, S. (1989). Electrochemically induced nuclear fusion of deuterium. *Journal of Electroanalytical Chemistry,* 261, 301–308.
[6] Asimov, I. (1965). *A Short History of Chemistry.* London, UK: Greenwood Press.
[7] Theobald, D. W (1968). *An introduction to the Philosophy of Science.* London, UK: Methuen.

Chapter 2

WHAT IS CHEMISTRY ABOUT?

We would like to start proposing an initial question: what does the study of chemistry involve? In other words, which specific "subject" does Chemistry want to deal with?

In attempting to answer these questions, we would like to quote a book by G. Bachelard[1] (1884–1962), entitled "Rational Materialism" [1]. This book examines the effort man has put into recognizing the hidden order of chemical features in the natural world, or the rational discovery of all relationships which are present in the mineral kingdom and the attention given to explaining them. These relationships are not often accessible and immediately comprehensible and one of the major impediments along the road to success was the fact that at first sight nature gives the impression of a great disorder. The task of a person involved in studying chemistry is to discover a rational organization which is able to make order out of the chaos among natural substances. However, nature[2] is a complex system, and before the publication of the Elements of Chemistry by A. Lavoisier (1743-1794), the simplest principles of chemistry had then hardly dawned. As G. Bachelard says:

---

[1] An essay about the philosophy of science of Gaston Bachelard written in English can be found in the book by M. Tiles [2].
[2] We suggest the reading of a really interesting article with a nice, thorough analysis of the notion of nature in chemistry by J. Schummer [3].

> the Earth, in its actual phenomena, is very poor of chemical teaching aids. Without any doubt we walk on sulphides and oxides and live in structures which are mainly made by carbonates [1] (our translation).

Hence it is far from being simple to understand chemical processes. If mineral deposits witness a chemical activity which had stopped millions of years ago, it is highly improbable they continue to show any chemical action. Surely are we witnesses of a certain chemical activity as, for example, in the eruption of an active volcano, the direct effect of atmospheric agents on rocks, soils and minerals, or the formation of stalactites and stalagmites from mineralized water solutions. However these phenomena are too complicated to be explained without a chemical rational apparatus. It is not easy to find a precursor of a discipline in the history of the sciences, because many people worked contemporarily to the same problems, but without any doubt the evolution of a science like Chemistry really started after the pioneering work of A. Lavoisier [4, 5]. As shown by his work, in which he introduced quantitative measurement to the study of chemical processes and was able to achieve the law of conservation of mass, a rational approach started when the natural word was abandoned (because of the presence of complex chemical processes) and chemists trained their minds to study simple chemical processes by means of laboratory activities. For example, to say that a salt forms as the product of the reaction between an acid and a base, it is necessary first to have convincing proof of the purity degree of acids and bases.

Now it is important to recall some of the steps that have marked the chemical "route", without dealing with all the historical details that have led to the development of chemistry, which can be easily found in many books [5, 6]. Some of these steps were crucial and involved the overcoming of psychological and technological obstacles: a) it was necessary to more clearly identify the difference between physical properties of matter, such as light, heat, electricity and chemical elements; b) even if the vast majority of elements were known, some of them were missing, which did not allow for a proper classification; c) the difficulty of finding new elements clashed with the recognition of their differing amounts in nature.

Now, which are the most important factors of this rational route?

If the first step was the abandonment of the natural world on the basis of the explicit recognition of natural substances not behaving simply, then analysis becomes the dominant concern in the following step. Extensive research effort was therefore devoted to decomposing simple and complex compounds in order to find their combining-ratios by weight [1]. It is important to remember the time when it was discovered that water was not an element, but a compound with a precise formula always obtained by the reaction of two parts of hydrogen with one part oxygen. The experiment done by H. Cavendish (1731-1810), in which he burned his inflammable gas (hydrogen) producing vapours that, when condensed yielded water, was lately interpreted by A. Lavoisier as a simple combustion reaction, namely a reaction of combination of hydrogen with oxygen [6].

Table 1. Sodium Chloride NaCl FW 58.44

Grade	ACS reagent[*]
Assay	$\geq 99.0\%$
total impurities	$\leq 0.005\%$ Insoluble matter
pH	5.0-9.0 (25 °C, 5% in solution)
Mp	801 °C (lit.)
anion traces	bromide (Br^-): $\leq 0.01\%$
	chlorate, nitrate (as NO_3^-): $\leq 0.003\%$
	iodide (I^-): $\leq 0.002\%$
	phosphate (PO_4^{3-}): ≤ 5 ppm
	sulfate (SO_4^{2-}): $\leq 0.004\%$
cation traces	Ba:, passes test
	Ca: $\leq 0.002\%$
	Fe: ≤ 2 ppm
	K: $\leq 0.005\%$
	Mg: $\leq 0.001\%$
	heavy metals (as Pb): ≤ 5 ppm

[*] it meets the specifications of American Chemical Society

Analysis was connected to the effort of obtaining purer and purer reagents. So as, if the first concern was therefore the analysis, the second became purification and, hence, the research for effective purification and separation methodologies and techniques [1], even if chemical practices as analysis and purification should be considered as interlacing. In fact, it is only when elements and simple compounds are available in several grades of purity (a real purity certification), that Chemistry as a real science had begun. It could be interesting to examine, just out of curiosity, a current product label of a well-known chemical compound like sodium chloride sold by Sigma-Aldrich, having an ACS reagent classification; the purity of which is well characterized and documented (see Table 1).

As we have shown above the scientific concern of chemistry is with classifying, systematizing, and attempting to understand the properties of materials and their transformations [7]. Therefore, this attempt of discovering rationality within the structure of matter and the problem of transformation of substances has continued by interpreting properties of matter making use of geometrical schemes. The German chemist F. A. Kekulé[3] (1829-96), following the suggestion of the Scottish chemist A. S. Couper (1831-92) who used to put small dashes among atoms, began to work out the formulas of many organic compounds, and this is the moment in which the first formulas appeared with their explicative function [6, 8]. In other words, the use of structural formulas became widespread. They were hypothetical at the beginning, but step by step they developed to provide a realistic description. If we look at a structural formula, it is not correct to state that it is only a conventional way of representing matter, because it goes without saying it would not have anything to do with reality. On the contrary, structural formulas are graphical representations which explain physical properties and suggest which chemical experiments may be carried out. When a formula is given, there are chemical experiments which are a priori forbidden because they are not consistent with it [9]. On the other hand, there are experiments which have been tried out on molecules on the basis of their structural formulas. In other words, the reactivity of a chemical compound can be predicted on the basis of its structural formula. This approach to the chemical problem demonstrated to be quite effective:

> *Structural formulas are the appropriate representation to make predictions of chemical properties, [and] it is the only way we have until today by which such predictions can be derived systematically* [10].

---

[3] Friedrich August Kekulé von Stradonitz, was one the of the fathers of Organic Chemistry, and, in particular, of the first structural hypothesis of the benzene molecule. His was a pioneering work in the use of structural formulas to predict physico-chemical properties and reactivity of organic compounds.

As shown in Figure 1, the diverse geometrical dispositions of carbon, hydrogen and oxygen atoms in molecules accounts for their different physical and chemical properties.

In fact, structural formulas, though representing:

> *highly stylized representations of very complex material systems [are grounded in experimental reality, thus by] interpreting reactivity and spectroscopic data of [pure compound we were able] to reconstruct very basic features of the "geometry" of those microscopic particles [which we call molecules]* [11].

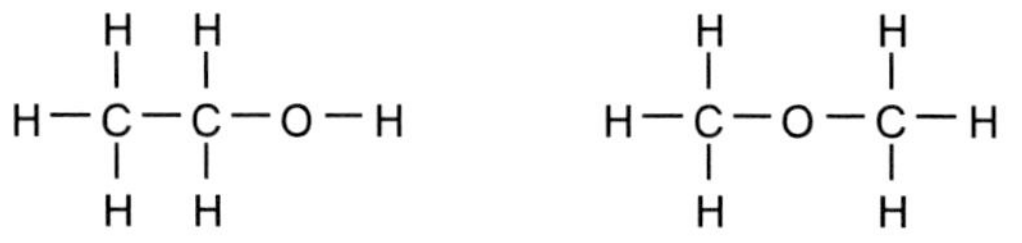

Figure 1. These compounds have different physical and chemical properties owing to their structural diversity, a) ethanol (left) and dimethyl ether (right), sharing the same chemical C_2H_6O formula, are liquid and gas respectively at standard conditions, b) maleic acid (left) and fumaric acid (right), sharing the same chemical formula $C_4H_4O_4$ show different solubility in water and display very different melting points, 135°C and 287°C, respectively.

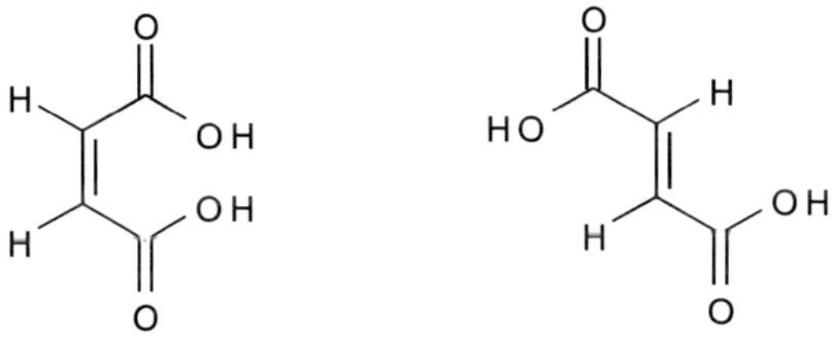

From this perspective, it could be useful to think back to the discovery of the benzene formula. We know that Kekulé hypothesized a hexagonal formula with three alternate double bonds, but this representation does not explain the benzene chemical behaviour, even if we considered the reality of the benzene molecule resonating between its two canonical forms. The benzene molecule is not an alkene, because it does not contain the unsaturated C=C functional groups which characteristically undergo addition and polymerization reactions. Today, the modern physico-chemical techniques for structural studies and the quantum methods for theoretical calculations have led us to understand with certainty how the benzene molecule is made and, hence, its chemical formula has to reflect its physical and chemical properties. The delocalized π bond (the molecular orbital description is certainly more complicated) over the six C-atoms of a planar benzene molecule, is simply symbolized by a circle inside the regular hexagon formula.

Is it sufficient to recognize that a rational organization of matter could be only achieved on the basis of structural theories? Or does the attempt to find out a rational explanation about chemical transformations need something else? The contemporary chemical science not only describes chemical phenomena, but also wants to give reasons for the occurrence of them in the material world. So as, one important thing to note is that the discovery of a rational order of matter was achieved by studying the energy exchanges during chemical reactions. In particular, as G. Bachelard puts it:

> *this rational materialism is an energetic materialism* [1] (our translation).

In fact, if we really want to understand the reasons behind matter transformation, we have to highlight and explain these energy exchanges, otherwise we would only obtain a mere description of chemical phenomena. Therefore, a chemical event is the demonstration of an energy, or rather the balance of energy between what has been destroyed (supplying energy) and what has been formed (gaining energy). Looking at it from this perspective, water electrolysis could be an interesting example to discuss.

Therefore, the knowledge of these energy exchanges becomes essential for the study of the reaction energetics. It will occupy a position secondary to that of the description of the chemical phenomena, but is an ineluctable necessity if one wants to deeply comprehend causes and reasons of why chemical events occur. Matter transformations can be accounted for according to energy laws. Disciplines like Chemical Thermodynamics (which will be discussed in detail in chapter 6) and Atomic and Molecular Spectroscopy have played an invaluable role to get to the root of the problems under chemical phenomena.

We ought to remark, before concluding the discussion of this chapter, how formulas, by "geometrizing" matter, played the role of capturing the chemist's imagination in conceiving possible new experiments. And while formulas are used in such a way to represent reality accurately, so are the study of energy exchanges of great importance for understanding the hidden reasons for the occurrence of chemical events.

REFERENCES

[1] Bachelard, G. (1972). *Le Matérialisme Rationnel*. Paris, France: Presses Universitaires de France (PUF).
[2] Tiles, M. (1984). Bachelard: Science and Objectivity. Cambridge, UK: Cambridge University Press.
[3] Schummer, J. (2003). The notion of nature in chemistry. *Studies in History and Philosophy of Science, 34*, 705-736.
[4] Mason, S. F. (1962). *A History of the Sciences*. New York, NY: Collier Books.
[5] Leicester, H. M. (1956). *The Historical Background of Chemistry*. New York, NY: John Wiley & Sons Ltd.
[6] Asimov, I. (1965). *A Short History of Chemistry*. London, UK: Greenwood Press.
[7] Christie, M., & Christie, J. (2000). *Laws and Theories in Chemistry Do Not Obey the Rules*. In N. Bhusham & S. Rosenfeld (Eds.), Of Minds and Molecules, New Philosophical Perspectives on Chemistry (pp. 34-50). New York, NY: Oxford University Press.
[8] Partington, J. R. (1957). *A Short History of Chemistry, London, UK: Macmillan,* reissued (1989) by New York, NY: Dover Publications.
[9] Bachelard, G. (1971). *Epistémologie. Textes choisis.* (essays collected by Lecourt, D.) Paris, France: Presses Universitaires de France (PUF).
[10] Schummer, J. (1998). The Chemical Core of Chemistry I: A Conceptual Approach. *HYLE - International Journal for Philosophy of Chemistry,* 4, 129-162.
[11] Tontini, A. (2004). On the limits of the Chemical Knowledge. HYLE *- International Journal for Philosophy of Chemistry*, 10, 23-46.

Chapter 3

ABOUT THE CONCEPT OF VALENCE

This third part, which introduces the valence concept and the notion of chemical bond, has been conceived as an attempt to give answers to questions that have changed over time, since advances in science have changed the directions of questioning towards a grounding of our understanding of physico-chemical reality.

The valence concept was developed when chemists begin to understand what combining ratios within a compound could mean and lately how atoms of elements could be spatially arranged around a central atom thus giving existence to individual molecular structures. Over time there have been several theories which have dealt with an indefinite concept of chemical bond, all of which have tried almost explicitly to answer the following question: how could the formation of a molecule be explained on the basis of the chemical properties of the atoms within its framework?

It is not easy to find clear attempts to address this problem nor is it easy to return to the search for precursors of a theory. The first reliable proposal was the Theory of Types by A. Laurent (1808-1853) [1, 2], in which it was stated that organic molecules (the difference between organic and inorganic compounds was not yet defined and organic chemistry was seen as a chemistry of substances belonging to natural world) could be classified into families (types) with similar characteristics. Molecules like methanol, ethanol or dimethyl or diethyl ethers could be ideally derived from water by substituting hydrogen atoms with one or two methyl/ethyl moieties, respectively (water type molecules). Aliphatic amines, which are primary, secondary or tertiary amines, can be considered as derived from ammonia by substitution of hydrogen atoms of ammonia with alkyl groups (ammonia type molecules). The smell of very small amines like methylamine and ethylamine is very similar to ammonia. Hydrocarbons could be conceived as derived from the hydrogen molecule by substituting a hydrogen atom with an alky group (hydrogen type molecules) and some derivatives could be considered as deriving from acids like hydrochloric acid, for example ethyl chloride was conceived as a hydrogen chloride type molecule.

These early theories, which may be considered as attempts to give order to known compounds, unfortunately lack the capability to explain the reasons behind molecule formation. However, these theories played the important role of pointing out regularities which were not considered or were judged only superficially before. For example, the oxygen atom always had two "entities" as hydrogen atoms (in water) or one hydrogen and an alkyl group (in alcohols) or two alkyl groups (in ethers), while in the same way the nitrogen atom always involved three groups. Analogously, by observing the dissolution of metals in acids (zinc always reacted with a double amount of hydrochloric acid) or looking at the reactivity of phosphorus or arsenic compounds with chlorine (three or five chlorine atoms in their compounds), the following idea was developed: regardless of the characteristics of the bound elements, the combining power of the attracting (central) element is always satisfied by the same number of atoms. This deduction allowed E. Frankland (1825-1899) to state his idea of valence, which was later named Valence Theory [1, 2]. This theory attributed, as a distinctive property of elements, valence, i.e. a combining power that was fixed or variable, to each element and tried to explain all the combination ratios obtained in the elemental analysis on the basis of the valence of the elements. It is important to understand that, in its simplicity, this theory gave reason to the most of empirical formulas and gave them meaning, because one or more valences could be ascribed to each element. In a way, Frankland's theory was a first attempt to address our initial question. Certainly, it could not go into the physical reasons of compound formation (why molecules as aggregates of two or more atoms are more stable than atoms themselves?), but it was very important in the development of Organic Chemistry.

Later on, F. A. Kekulé affirmed as a *postulate* that carbon is always tetravalent, a cognition a priori from which Organic Chemistry stemmed [3, 4]. Then as stated above, a short segment was introduced to explain how an atom is linked to another in a molecule [2], an initial idea of the chemical bond, which we discuss in more detail later on. The Kekulé's formulas were magnificent answers to many of the Organic Chemistry's problems, also tackling that of geometrical isomerism (geometrical isomers have the same crude chemical formula, but are structurally dissimilar and have different physico-chemical properties like melting point, boiling point, viscosity, density and chemical reactivity) up to the hypothesis of the hexagonal formula of the benzene molecule.

There are several points along this path of reasoning which need to be stressed:

1. On the basis of this simple model, the enormous work of the synthesis of organic compounds started.
2. The Organic Chemistry example was that of doing science by means of a non-mathematical type of logical reasoning, which has had a great success. In other words, the growth of Organic Chemistry [3] was the example of a non-quantitative type of logic reasoning, a way of looking at reality not driven by the strict mathematical logic, but marked by an enormous success.
3. A philosophical dispute arose among chemists, between supporters of the element fixed valence (organic chemists) and those of the variable valence (inorganic and analytical chemists).[1]

---

[1] Even though we do not agree with Feyerabend's criticism [5] to the notion of scientific progress, we have found the category of violation of a scientific procedure appealing enough to us. Many scientific advances have not been achieved according to a rational scheme and methodological criteria, and rules have often been infringed, showing scientists to be real opportunists. In fact, they do not disdain using any means, methods and materials to accurately describe and make predictions within the little piece of the world to which they are accustomed. Based on this, it could be a little rash to talk about a unique scientific tradition, because science consists of peculiar cognitive domains which are irreducible to one another. This does not mean scientific research is an arbitrary procedure or proceeds blindly, but its criteria are restricted to the specific research process and do not come from abstract theories about rationality. Violations of the supposed scientific logic have been often repeated in view of the success of a certain research project (call it scientific progress). The growth of Organic Chemistry has been substantially based on the Frankland's valence concept, i.e. the combining power of a given element. Now, while many chemists found a variable valence in most of their analyses of inorganic compounds (an element within its diverse compounds would have shown different combination ratios), organic chemists assumed that carbon had a fixed valence

The last breakthrough, which came from the valence theory, was the discovery of optical isomers [1, 2], compounds having the same physico-chemical properties except the effect of rotating differently the polarized light plane. J. A. Le Bel (1847-1930) and J. H. Van't Hoff (1852-1911) hypothesized the four valence of carbon tetrahedrally directed in a three-dimensional space, thus initiating the tradition of stereochemistry. As we know, two optical isomers differ from one another on the grounds of the spatial arrangement of their atoms around a central carbon (which is named asymmetrical) and are mirror images of each other.

However, this valence concept is revealed to be sufficiently artificial as it did not go deeply into the real nature of the chemical bond. In other words, it could not be the answer to the following question: what kinds of force play a relevant role in a molecule formation, starting from the characteristics of the atoms which are bound in the molecule?

The development of Physical-Chemistry in its fundamental fields, Gas Kinetic Theory, Chemical Thermodynamics, Electrochemistry, Molecular Spectroscopy and Chemical-Bond Theories, allowed to find a complete answer to this last question. A new object came into play: the particle called electron was discovered. The first studies in the electrical conductivity of saline solutions by S. A. Arrhenius (1859–1927) [2] demonstrated that some compounds are able to behave as electricity conductors, because salts dissociate into ions, which are freely moving charged particles able to conduct electricity. Salts give off ions when dissolved in water, and, thus they are called electrolytes, but not all compounds behave this way. Two new words were coined by chemists to identify chemical compounds on the grounds of their chemical bond, electrovalence (an ionic bond) and covalence (a covalent bond). The former explains the formation of ionic crystals, in which the physical force acting within is electrical, meaning that the attraction between particles comes from opposite electrical charges. The latter accounts for valence

---

(the carbon atom invariably has a four-valence) and, therefore, they became supporters of the fixed-valence "party". Meanwhile, starting from the latter assumption, which is an obvious contradiction of experimental findings (i.e. the valence of an element could be variable), organic chemists started to write formulas of the majority of organic compounds. These formulas were able to provide reasonable and accurate predictions of physical properties and chemical reactivity of the majority of organic compounds and helped enormously in the development of Organic Chemistry. Without any doubt, the acceptance of organic formulas was also connected to dye manufacturing as well as the expansion of the German chemical industry [6].

electrons being localized in particular spatial regions of a molecule in order to minimize repulsive forces among nuclei of its atoms and is responsible for the formation of molecular compounds. In the simplest cases, the electronic configurations of atoms belonging to a molecule or to ions seemed to obey to the octet rule that is the obtainment of a stable eight-electron configuration similar to that of noble gases. These are obviously two extreme cases, because the presence of both covalent and electrovalent characters does actually exist for the most of compounds.

It is important to note that despite having conserved the "valence" term, these first modern attempts have nothing to do with Frankland's theory of valence, since they began with giving a physical rationale for the initial question and gradually the expression "Chemical Bond Theory" claimed everyone's attention. This tendency to save an old scientific name adding a new meaning to it is a typical scientific habit. It could be interesting to observe the development of the symbolism of the chemical bond. At the beginning the line representing the bond in a structural formula only indicated the element valence, but now the same line acquires the significance of Lewis' electron pair (G. N. Lewis, 1875-1946):

> *the line between two atoms in a molecule has acquired meaning and importance* [3, 4] (our translation).

The advent of quantum mechanics marked a new epoch in the study of the chemical bond. Heisenberg's Uncertainty Principle (W. Heisenberg, 1901-1976) changed the way of thinking of atoms and molecules. If it is impossible to identify the precise localization of electrons, we are constrained to open up to another way of thinking, in which orbitals, volumes, electronic densities, symmetries, the principle of maximum overlap and so on are to be considered. The mathematical architecture of quantum mechanics is vast and complex, and we advise the reader to consult specialized books. What is interesting to note here is that quantum chemistry often seems to consider that the concept of the chemical bond is useless, as explicitly asserted by W. Kutzelnigg (1933-), a leading German theoretical chemist, in the introduction of one of his articles:

> *There is nevertheless an extremely wide gap between descriptive theoretical chemistry used in chemistry textbooks and computational quantum chemistry of present-day research. Many concepts from early days*

> *of quantum chemistry that have found their way into descriptive theory have become insignificant or useless in current quantum-chemical research. It often seems that even the chemical bond itself is one such concept. Ideas about binding patterns are not necessary input for calculation of the properties of a molecule from approximate solutions of the Schrödinger equation. Such ideas become relevant only when one attempts to gain a physical understanding of the results of the calculation, or even to make significant statements about the molecule without calculations. [...] The fact that a question is asked does not mean that its answer is known. The question of the nature of the chemical bond has by no means been fully clarified* [7].

These problematic assertions by Kutzelnigg[2] are not absolute and have to be looked at proportionally. This is shown by R. J. Gillespie and I. Hargittai in chapter 7 of their book on the molecular geometries [11], in which they showed, on the basis of the theoretical work by R. F. W. Bader (1931-), that definite bond paths can be actually inferred by determining the differential electronic density maps of chemical compounds. R. F. W. Bader himself showed in recent articles [12,13] how atoms within a

---

[2] These assertions raise a set of appropriate questions about the interconnection between physics and chemistry. Should chemistry be considered an appendix of quantum mechanics? Or rather can quantum mechanics explain all experimental chemical events? Can we really assert that all the characteristics of chemical compounds could be grouped into a wave function, which is able to make correct previsions? It is important to state that chemistry has a proper methodological and ontological autonomy, and thus, it remains autonomous in its development with respect to other natural sciences [6]. As E.R. Scerri has shown in his book [8] on the periodic table of the elements and also elsewhere [9], many of the chemical properties of the elements are not derived from quantum mechanics' atomic theory, but they have been obtained from the chemical behaviour of elements and compounds and from spectral observations. Furthermore, to obtain the ground state electronic configurations of isolated atoms, Hund's rule and Pauli's principle are at the same time mere empirical findings and/or causal explanations *ad hoc* in order to get a satisfactory agreement with experimental observations. In the same way, one could ask if the notion of chemical bond can be reduced to a simple or complex quantum mechanics description only. Particular electron configurations have to be considered to satisfy the current notion of chemical bond and they are derived from spectroscopic measurements, not from basic principles of quantum mechanics [6]. Concepts related to the molecular structure such as chemical bond, valence of elements, molecular conformation and so on, did not derive historically from mechanics equations or from a general Schrödinger equation for molecules [10]. Unfortunately, many of our colleagues continue to think that chemistry can be reduced to physics as many of them teach that chemical bond notion can be reduced to quantum mechanics. Not only they pretend people have to abandon their rational and logical schemes used in the chemical practice, but also they want them to assume the point of view of those working as physico-chemical theoreticians. As a joke we often address this dogmatic attitude: at the beginning God created the heavens and the earth and even the quantum mechanics (or the Schrödinger equation) from which Chemistry derived!

molecule are connected by particular bond paths, through the topology determination of the electron density.

But, now let us focus on the last of Kutzelnigg's assertions by saying that if a question is asked, its answer should exist. Human beings are not yet truly aware of the way to answer all the questions that are asked, but the certainty of finding an answer in response to the question put forward is the drive which pushes scientific research; it is the certainty that motivates scientists to search for a way to achieve it.

We would like to conclude this third section by citing a statement taken from Einstein's (A. Einstein, 1879-1955) reflections upon the connection between reality and its intelligibility:

> *The very fact that the totality of our sense experiences is such that by means of thinking [...] it can be put in order, this fact is one which leaves us in awe [...] One may say "the eternal mystery of the world is its comprehensibility" [...]. It implies: the production of some sort of order among sense impressions, this order being produced by creation of general concepts, relations between concepts, and by relations between concepts and sense experience [...]. It is in this sense that the world of our sense experiences is comprehensible [...]. The fact that it is comprehensible is a miracle* [14].

REFERENCES

[1] Leicester, H. M. (1956). *The Historical Background of Chemistry.* New York, NY: John Wiley & Sons Ltd.

[2] Asimov, I. (1965). *A Short History of Chemistry.* London, UK: Greenwood Press.

[3] Bachelard, G. (1972). *Le Matérialisme Rationnel.* Paris, France: Presses Universitaires de France (PUF).

[4] Bachelard, G. (1973). *Le Pluralisme Cohérent de la Chimie Moderne,* Paris, France: Libraire Philosophique J. Vrin.

[5] Feyerabend, P. K. (1988). *Against Method. Outline of an Anarchistic Theory of Knowledge.* London, UK: Verso.

[6] van Brakel, J. (2000). *Philosophy of Chemistry.,* Leuven, NL: Leuven University Press..

[7] Kutzelnigg, W. (1973). The Physical Mechanism of the Chemical Bond. *Angewandte Chemie International Edition,* 12, 546-562.

[8] Scerri, E. R. (2007). *The Periodic Table, Its Story and Its Significance.* New York, NY: Oxford University Press.

[9] Scerri, E. R. & McIntyre, L. (1997).The Case for the Philosophy of Chemistry. *Synthese,* 111, 213-232.

[10] Del Re, G. (1996). The Specificity of Chemistry and the Philosophy of Science. In V. Mosini (Ed.), *Philosophers in the Laboratory* (pp. 11-20). Rome, Italy: Musis, EUROMA.

[11] Gillespie, R. J., & Hargittai, I. (1991). *The VSEPR Model of Molecular Geometry. Boston*, MA: Allyn & Bacon.

[12] Bader, R. F. W. (2011). On the Non-existence of Parallel Universe in Chemistry. *Foundations of Chemistry.,* 13, 11-37.

[13] Bader, R. F. W. (2010). Definition of Molecular Structure: By Choice or by Appeal to Observation? *Journal of Physical Chemistry A,* 114, 7431-7444.

[14] Einstein, A. (1950). *Out of my later years.* London, UK: Thames and Hudson.

Chapter 4

ABOUT THE PERIODIC TABLE

This fourth part is centred around the Mendeleev's periodic table, but we do not hesitate to go into the modern periodic system and aim to help students understand the differences that arise from different approaches to the same scientific problem. If the reader is interested in this subject, we suggest the reading of the book by. E. R. Scerri [1] on the same subject, as it is an almost comprehensive and very detailed account of how things have proceeded over the last two centuries. An interesting biography of Mendeleev, which includes personal and familial aspects of his life, was published recently [2].

It is important to recall some of psychological and technological obstacles to conducting systematic and harmonic studies of simple substances[1] as mentioned above.

We can summarize them in four points:

1. The vast majority of elements were known, but some of them were missing, not allowing for any reliable attempt at classification.

---

[1] This term "simple substance" needs clarification. This question arises: what is the correct meaning of the term "simple or pure substance" [3]? Is a simple substance to be defined in terms of its composition or molecular structure (namely, invoking its microstructural models on the basis of a theoretical approach) or should we consider its physico-chemical properties as density, refracting index, boiling and melting points and so on? '*Historically, the concept of substance and element were developed while interacting with the notion of molecule and atom [...]. Any talk of atoms, molecules, and valences will be relative to this macroscopic scientific definition of pure substance. Notions like temperature, density, phase conversion and so forth, are used to give a definition of pure substance within the scientific image, against the background of the thermodynamics*'[4].

2. Physical properties of matter such as light, heat, electricity, and so on, were considered as simple substances; in other words the distinctive characteristics of physical objects interfered with the definition of a pure substance, elements or compounds, all of which have a definite composition.
3. The difference in the naturally occurring amounts of elements has made their identification difficult. In order to organize and classify elements or, in other words, to achieve a systematic and rational approach to periodicity, it was necessary to attribute no significance to the idea of quantity; in fact, at a rational level, for example, an element of the rare earths (even if, around 1850, the known number of rare earth elements were at least six) plays its role like any the most of abundant elements.
4. It was necessary to get rid of the tendency to explain chemical facts by making analogies with the natural world, namely turning to a more general phenomenon, perhaps more immediate. We will see that Chemistry progressed when it tried to differentiate (i.e. to seek its own explanation) rather than to search for a generalization of immediate empirical data [5].

Let us start with the following initial question: from what did chemists start to recognize a rationale in the organization of simple substances?

As almost always happens in our life, everything starts from the previously acquired knowledge we had. Therefore, it is clear that a limited knowledge renders a rational synthesis very difficult. For the first attempt to recognize the hidden order of simple substances we had to wait up to the second half of the 19th century. This historical period is dominated by the figure of the Russian chemist D. I. Mendeleev (1834-1907), even if other attempts to look for regularities among simple substances had been made before him. Two interesting attempts must be mentioned, the first by J. A. R. Newlands (1837-1898), who realized that elements with similar properties tended to assemble themselves in the same horizontal line [6, 7], when arranging them in order of increasing atomic mass (actually, earlier chemists considered atomic weights, but today it is better to talk of atomic mass).[2] The second by A. Béguyer de Chancourtois (1820-1886) [6, 7],

---

[2] The notion of atomic weight needs a further clarification, and following the analysis of G. Bachelard [8], we can state that, in its first determination during the 19th century, the atomic

who built a cylindrical diagram in which elements having similar properties lined up vertically, in order of increasing atomic mass.

A system contrary to Mendeleev's periodic table was that developed by J. L. Meyer (1830-1895), who took the atomic volume into consideration. He considered the atomic volume to be an elemental property and obtained it from the ratio between atomic masses and densities [8]. If the piece of information given by the atomic volume is reported as function of increasing atomic mass, a curve with peaks corresponding to alkali metals was its graphical form, as shown in figure 1.

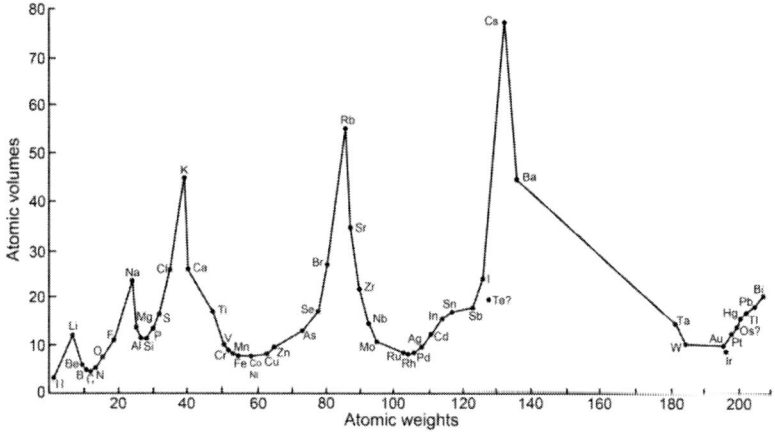

Figure 1. Meyer's plot of atomic volumes of elements against their atomic weights.

Unfortunately, the details of Meyer's discovery were reported a year later after those of Mendeleev' table, so this interesting system was never taken into consideration and remained an old scientific curiosity. In any case, all these initial systems had a common ground: they tried to arrange elements in order of increasing atomic mass. The knowledge of atomic masses (even if they were not always determined with sufficient accuracy)

---

weight was thought to be a relative weight, because it was a ratio between two weights (that of the proper element and that of the element chosen as unit). In other words, its significance derived from the combining ratios between weights of reagents in a chemical reaction, and therefore, it was thought as a relative number. After the work of J. B. Perrin, 1870-1942, [9] who was able to compute Avogadro's number, given the number of atoms or molecules contained in a mole of elements or compounds, the atomic weight became an absolute number, because it refers to the average mass found averaging over natural abundances of the isotopes of each element.

and the valence were the data from which Mendeleev started. A reaction stoichiometry always gives molar ratios among reagents and products from which atomic masses and valences of elements were obtained. The importance of the Mendeleev's periodic system was the organization of the elements in families and, later on, the unique system that these families comprised. As a first ordinal pattern, he arranged the elements in order of increasing atomic masses. As a second ordinal pattern he considered their valence. In other words, he wrote known elements in a column as a function of their increasing masses, but interrupted the column, while continuing to respect the sequence of atomic masses, to put the elements which presented the same valence (actually, the same chemical characteristics) in the same horizontal line. The important thing to point out is the ordinal character of these two variables [5, 8].

Let us imagine the final table taking shape. Table 2 shows how a probable Mendeleev's periodic system, comprising the first thirty-four elements with atomic masses up to 87.6, could have been arranged.

Table 1. A simplified sketch of the Mendeleev's periodic system up to the element having 87.6 atomic mass

				21° Ti = 50.0
				22° V = 51.0
				23° Cr = 52
				22° Mn = 55
				24 Fe = 56
				25° Ni = Co = 59
1° H = 1				26° Cu = 63.4
	3° Be = 9.4	10° Mg = 24.0		27° Zn = 65.2
	4° B = 11.0	11° Al = 27.4		28° --- ★
	5° C = 12.0	12° Si = 28.0		29° --- ★
	6° N = 14.0	13° P = 31.0		30° As = 75.0
	7° O = 16.0	14° S = 32.0		31° Se = 79.4
	8° F = 19.0	15° Cl = 35.5		32° Br = 80.0
2° Li = 7.0	9° Na = 23.0	16° K = 39.0		33° Rb = 85.4
		17° Ca = 40.0		34° Sr = 87.6
		18° --- ★		
		19° ?Er = 56		
		20° ?Yt = 60		

As can be seen from Table 1, Mendeleev did not know the existence of helium (nor the other noble gases which are not present in his table); the first evidence of helium and the other noble gases came in about the 1869, the same year in which the first version of his periodic system was presented to the scientific community.

We think it is important to stress that sometimes he was constrained to contradict the atomic mass sequence, by arranging elements having the same valence in the same horizontal line. As it is shown in Table 2, tellurium (AM = 128.0) came before of iodine (AM = 127.0).

Mendeleev did not think of atomic masses as approximately determined, but made a choice in favour of the element valence on the grounds of his chemical sensibility: he considered the valence of elements a property more important than their atomic mass.[3]

Table 2. A simplified sketch of Mendeleev's periodic system from the element with 118 atomic mass up to the element with 128

C	Si		49° Sn = 118	Tetravalent
N	P	As	50° Sb = 122	Trivalent
O	S	Se	51° Te = 128	Divalent
F	Cl	Br	52° I = 127	Monovalent

Mendeleev not only produced the first plausible periodic system, but also he left some vacant spaces (see stars in Table 1) and, instead of considering them as imperfections of his table, predicted the possibility of the existence of new elements not yet discovered. In many respects, he was ahead of his time, and after a few years, gallium (Ea, eka-alluminium), germanium (Es, eka-silicon) were promptly discovered [11]. Many of the properties of scandium, discovered in 1879, and technetium, discovered in the late 1930's, were also predicted by him and the elements indicated with the provisional names of Eb, eka-boron and Em, eka-manganese,

---

[3] E. R. Scerri seems to agree with this last assertion in a previous article [10], but in his book on the Periodic Table [1] he questioned it, pointing out that valence was not the only important criterion in Mendeeelv's point of view, because many elements showed variable valence, and that he preferred to concentrate his attention on the ontological concept of a basic substance (or abstract element), which surely is an abstraction or a transcendental concept which guided him to make precise choices.

respectively [6]. It is important to recognize that experimenters set out to search for what he had indicated.

What is the limit of Mendeleev's periodic system? While it is true that this first version of the periodic table allows one to recognize a series of regularities in the chemical properties of elements, however it does not explain why these regularities exist. In fact, soon after Mendeleev' presentation of his periodic table, the question changed: why do regularities in the chemical properties of elements exist?

Only the contribution of atomic theory was able to answer this question.[4] The chemical behaviour of simple substances was studied in a better way through a rational theory, dealing with chemical aspects. The electron, a new object of the atomic theory, as was able to account for the chemical periodicity. Then, a new abstract concept like that of atomic number took precedence over the atomic mass notion. Now, the new organizing variable becomes the atomic number, or the total number of electrons of an atom. Therefore, in our modern periodic system the location of an atom in the table is due to its atomic number. Earlier on, in the Mendeleev's table, the atomic number was the ordinal number, or the number that fixed the element place in the table (first, second, third, fourth and so on). In light of atomic theory this number, losing its abstractness, became a number which effectively counted the number of electrons (or protons) contained in an atom. This new finding brought about a new change and a complete revolution in the chemical knowledge: the simple ordinal number which fixed the place of an element in the Mendeleev's periodic system had become of primary importance:

> *from ordinal had become a cardinal number and it was as if the pagination of a book could account for the project which the book itself was written for* [5, 8] (our translation).

Therefore, our modern periodic system of elements has nothing to do with Mendeleev's system. In the latter, the basic periods signified facts

---

[4] We agree with E. R. Scerri [1, 10] who has also questioned this assertion because he thinks quantum-mechanical atomic theory was not able to adequately account for grouping some elements into their proper groups. In fact, this classification was made primarily on the basis of the chemical properties of elements. But, with students facing their first introductory chemistry course, we would not be able to account for all the problems which still exist in the attempt to comprehend the physical world and are surely a challenge to future research activities.

without explanation, which could also be seen as occasional coincidences due to the periodic variability of qualities, often not really characterized and not well numerically determined. When the valence concept was examined on the basis of atomic theory, empirical investigation resulted to have been done against a basic cognitive framework, making the chemist able to ascertain experimental facts, but not to explain them. Hence, the atomic theory plays the important role of a cultural foundation which gives reasons for the chemical behaviour of elements.

How is it that Mendeleev has been able to build up a periodic system similar to that which we know today, without any knowledge of atomic theory?[5] This happened because there is an imperfect parallelism between the increase in atomic masses and the increase in atomic numbers as shown by the Moseley's studies (H. G. J. Moseley, 1887-1915), on the X-ray emission from anticathodes of metallic substances bombarded by cathodic rays of sufficient energy [12]. Figure 2 presents plots of $(1/\lambda)^2$ of the K_α lines experimentally obtained for various elements versus their atomic numbers or atomic weights.

Now atomic theory not only explains the order of the elements provided by Mendeleev's periodic system, but also the existence of transuranic elements. It is difficult to understand why the number of elements might stop at uranium, in the presence of empty orbitals. The nuclear physics and the development of nuclear technologies have stimulated the discovery of new elements, precisely the transuranic elements. In order to explain this, the modern periodic system has to be revised at the nuclear level and, thus, it is necessary to think of notion of atomic number again. The atomic number represents the number of electrons of an atom as well as the number of protons contained in the atomic nucleus of an element. As it turned out, the energy of the X-ray emission from anticathode depended upon the nuclear charge, as Moseley had shown through a straight-line relationship between the inverse square law of the wavelength of the emitted X-rays and the atomic number. As asserted by E. R. Scerri,

---

[5] This question was posed to us by a student once when we were about to conclude this argument. At first we were really surprised, but after due deliberation, we came to the conclusion that his question was legitimate and needed an answer.

Atomic number is [...] fundamental in an ontological sense of referring to a deeper level of reality, namely the nucleus.[13]

Namely, the periodicity found in the chemical characteristics of the elements depends on:

Z = number of protons (atomic number)
A = number of protons + number of neutrons (mass number)
$A - Z$ = number of neutrons

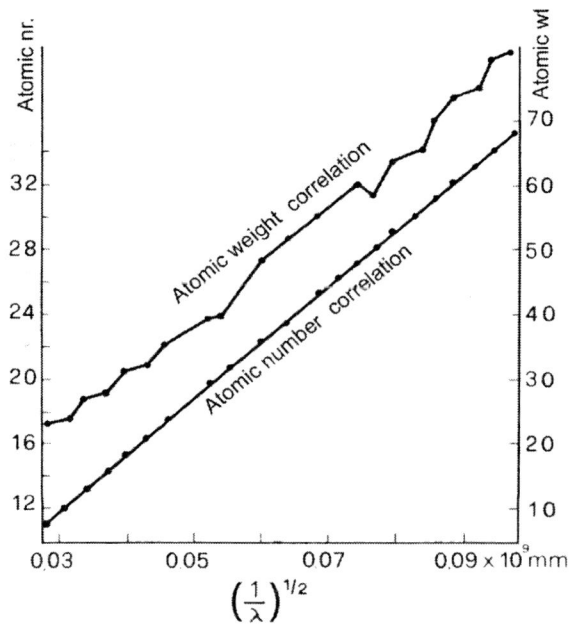

Figure 2. Plots of $(1/\lambda)^2$ of the K_α lines obtained from X-ray measurements on different anticathodes as a function of the atomic number or the atomic weight increases. It is possible to observe the perfect correlation with atomic number.

with these definitions in mind it is possible to know matter from A to Z [8] (our translation).

Now let us move on to an example taken from G. Bachelard's work [8]. Nuclear physicists are often said to be the scientists who realized the old dream of alchemists: to transform lead into silver or gold. However, alchemists tried to carry out their task with the primary goal of changing the qualities of matter, such as like colour:

> *If we could make the lead not only become yellow, but also if we could arrange to make lead heavier, as heavy as gold is, we could realize transmutation"* [8] (our translation).

As a matter of facts, on the basis of what we have already said and from a careful examination of the periodic table, this project would have led them in the opposite direction, making lead lighter rather than heavier.

Matter is organized in three levels:

a) the level of physical experiences;
b) the level of chemical experiences;
c) the level of nuclear experiences.

For example, the notion of density belongs to the first level and is not involved in chemical reactions. For the formation of a general compound, AB, what is important to know is the experimental conditions in which the elements, A and B, can react, but it is not relevant to know whether A has a lower or larger density than B. The task of alchemists cannot be carried out at the level of chemical experiences, because chemistry takes for granted the existence of different elements and does not take any interest about their transmutation, but rather about their interactions. In order to fulfil the dream of alchemists one would have to deal with the Z value (atomic number) of the lead nucleus, thus, the $Z = 82$ of lead should decrease to $Z = 79$ of gold. If the nuclear technology were able to take away three protons from the lead nucleus, it would have been possible to carry out the project, because all the other (physical and chemical) properties would have been surplus. If we cannot work at the level of nuclear experiences (namely, changing the number of nucleus protons), the hope to carry out a transmutation experiment is entirely in vain.

REFERENCES

[1] Scerri, E. R. (2007). The Periodic Table, Its Story and Its Significance. New York, NY: Oxford University Press.
[2] Woods, G.T. (2010). Mendeleev, the man and his matrix: Dmitri Mendeleev, aspects of his life and work: was he a some fortunate man? Foundations of Chemistry, 12, 171-186.
[3] Earley, J. E. (2009). How Chemistry shifts horizons: elements, substance, and the essential?. Foundations of Chemistry, 11, 65-77.
[4] van Brakel, J. (2000). Philosophy of Chemistry. Leuven, NL: Leuven University Press.
[5] Bachelard, G. (1973). Le Pluralisme Cohérent de la Chimie Moderne, Paris, France: Libraire Philosophique J. Vrin.
[6] Leicester, H. M. (1956). The Historical Background of Chemistry. New York, NY: John Wiley & Sons Ltd.
[7] Asimov, I. (1965). A Short History of Chemistry. London, UK: Greenwood Press.
[8] Bachelard, G. (1972). Le Matérialisme Rationnel. Paris, France: Presses Universitaires de France (PUF).
[9] Becker, P. (2001). History and progress in the accurate determination of the Avogadro constant. Reports on Progress in Physics, 64, 1945-2008.
[10] Scerri, E. R. (2000). Realism, Reduction, and the "Intermediate Position". In N. Bhusham & S. Rosenfeld (Eds.), Of Minds and Molecules, New Philosophical Perspectives on Chemistry (pp. 51-72). New York, NY: Oxford University Press.
[11] Atkins, P., & Jones, L. (1997). Chemistry, Molecules, Matter, and Change. New York, NY: W. H. Freeman and Company.
[12] Heilbron, J. L. (1966). The Work of H. G. J. Moseley. Isis, 57, 336-364.
[13] Scerri, E. R. (2010). Explaining the periodic table, and the role of chemical triads. Foundations of Chemistry, 12, 69-83.

Chapter 5

ABOUT ATOMIC THEORY

This fifth part contains some concepts to understand, which are of philosophical nature, and the reader should be acquainted to some of atomic theory notions such as those of wave-particle duality, Heisenberg's uncertainty principle, Schröedinger's equation, wave-function, and so on. We suggest addressing them to students, who are starting course in Advanced Inorganic Chemistry or Physical-Chemistry, courses in which the atomic theory is surely taught in an in-depth manner. Very often we find that when these topics are taught, the necessary scientific details are only addressed, without giving the students the opportunity to enter into the cultural aspects that are the key of the entire subject. Therefore, rather than going into the details of atomic theory, we will comment on some complex problems of cultural nature and give some general considerations about the daily work of scientists.

It is well known that 20th century science has been referred to as a new scientific revolution, in that it has upset what we had come to accept as fact on the basis of previous science outlook. It would not be incorrect to speak of the development of a new cultural position towards reality during this period. We can summarize this in the following points:

1. Modern science is no longer interested into things of our ordinary experience, as the relationship between scientific theory and practice has been overturned [1-3]. Scientific phenomena no longer involve things of our ordinary experience like a pendulum that swings or a heavy body that falls. Science nowadays is able to give

a detailed and subtle picture of the microscopical structure of an atom, but is so far from immediate human experience that sometimes it is very hard to understand how it was deduced.

> *The 19th century science [...] presented itself as a homogeneous knowledge; it presented itself as the science of our own world, in touch with our daily experiences organized by a universal and stable reason, and the final approval of which was our common interest. The scientist was [...], "one of us." He lived in our reality, he handled the same objects we did, he let the same phenomenon that educates us edify him, and he found evidence in the clearness of our very intuitions. He unfolded his demonstrations by following our geometry as well as our mechanistic physics. He did not bother with the principles underlying measurements, and he let the mathematicians worry about the game of axioms. Naturally, he, like us, embraced the same arithmetic. Science and philosophy spoke the same language [...]. Scientific hypotheses were considered schematic or pedagogical representations [...]. We were used to repeat that they were simple expressive means. Science, it was believed, was real through its objects and hypothetical by the bonds that were established between its objects. These hypotheses of bonds would be immediately abandoned should a most banal contradiction occur or should the tiniest difficulty surface during an experiment. We labelled these hypotheses conventional as if a scientific convention had no other means to be objective than through the rational.*
>
> *Thus, the new physicist has turned upside down the hypothetical perspective [...]. On the one hand, metaphors now represent the objects; and the way they are organized act as the real. In other words, what is now hypothetical is our phenomenon because our immediate take on reality only acts as a confused, provisional, and conventional given* [1, 4].[1]

2. The scientific object loses its uniqueness and the properties of a material substance. To better explain this assertion we would state the elementary particle idea must be looked at again. Actually, in the infinitely small the conventional idea of a body having mass, shape, durability, colour, position, speed, and so on, needs to be rethought, because a particle has completely different properties.

---

[1] This English translation of the text of G. Bachelard (which can be found in *Noumenon and Microphysics* and is one of the essays collected by D. Lecourt) was made by B. Roy [5].

- The elementary particle has neither specific dimension nor shape.

  Absolute sizes were not assigned to elementary particles, which are allocated in a region of influence (rather than a region of existence). The elementary particle exists in the place where it acts. The order of magnitude of a particle like an electron, which is about 10^{-13} cm, has to be interpreted as the shortest distance at which two of them can approach each other, when they interact with speeds near to the velocity of light. Therefore, the particle dimensions are dynamically defined, contradicting the idea that all bodies are impenetrable and, as a consequence, it means that an elementary particle does not possess specific geometrical forms. Geometry is an issue which starts to appear with molecules (molecular structures), that is to say, when more than two atoms react to form stable atomic aggregates.

  The element handed over to its loneliness has no geometry [1, 3] (our translation).

- The elementary particles are studied through their interactions.

  From the preceding assertion, we know that elementary particles can be studied only by observing what happens when they interact with each other. There is no other way to gain information on their properties.

  The elementary particle has neither position nor speed that can be defined simultaneously.

  Heisenberg's uncertainty principle sets a fundamental limit to the simultaneous knowledge of physical properties, such as position and momentum, which cannot be simultaneously measured with arbitrarily high precision (they behave in inverse relation, that is to say, if we desire high accuracy in the determination of an electron position we should give up determining its precise speed and vice versa). At an elementary level and in a first approximation, we address the consequences of Heisenberg's uncertainty principle by saying that if one wants to know about the energy (energy or speed or momentum are all interconnected quantities) of atomic levels, which are related to our observable events (i.e. atomic or molecular spectra coming from the interaction between light and matter), one has already made a choice: it is impossible to precisely state the electron location, which implies a probabilistic nature of its knowledge, so that we ought to think of it as spread around the nucleus. Some considerations about the philosophical interpretation of Heisenberg's uncertainty principle have been added in the foot-note.[2]

- The elementary particle is not an individual object.

An elementary particle does not have any distinctive trait. There is nothing which differentiates one electron from any other electron. This is why it loses its uniqueness.

- The elementary particle can be created or annihilated.

The law of mass conservation, which is the basis for quantitative relationships in chemical reactions, is valid when considering atoms or molecules (pure substances which have definite composition and constant properties), but loses its validity when elementary particles are considered: if an electron collides with a positron (a positive electron) they annihilate each other and production of electromagnetic radiation can be observed (two photons with 0.51 MeV each are produced). Similarly, if the interaction of electromagnetic

---

[2] Thanks to Heisenberg's uncertainty principle, the recognition of a prevailing theoretical limitation on the perfectibility of instruments and observations, did not threaten the trust in a complete physical determinism. In other words, science continued to give a picture of reality, probably an approximate understanding of it with respect to its goal, but a knowledge. Until Heisenberg's uncertainty principle was only considered a limitation to the precision of instruments, it was still possible to keep the notion of chance as opposed to that of cause. Unfortunately, it was soon given a drastic philosophical meaning. Something entirely different is meant if one follows the interpretation which implies that "what cannot be measured exactly cannot take place exactly". Hence, once granted a philosophical guise to the inability of science to measure characteristics of reality with absolute precision, the inability to grasp all the aspects of reality has been ascertained with philosophical accuracy. Then, chance takes the place of reality, and this is the reason why the question '*Could chance be considered real?*' is never raised in quantum mechanics. All this did not have any effect on the work of scientists who have proceeded with a great effort, so that science continues to be one of the highest values of our society, thanks to its ability to give information on the reality of matter and to allow for its manipulation. The only consequence was a crisis concerning the belief in a reality and any attempt at truth seeking. These considerations drew inspiration from the reading of an essay by S. L. Jaki (1924–2009) [6].

radiation with matter occurs and the incident photons have energies larger than 1.02 MeV, the radiation disappears and the creation of two elementary particles (a pair of an electron and a positron) can be observed.
- The elementary particle is only defined by its properties.

It is not possible to conceive of an elementary particle as separate from its properties. We cannot conceive of a photon if it does not have the velocity of light or an electron if it does not have a negative charge.

> *It surely seems that in the word of microphysics, the single object loses its substantial properties. There are then substantial properties only above - and not below - microscopic objects. The substance of the infinity small is of the same age as that of the relation* [1-5].

Before asking the main question, let us put forward some considerations about the role of scientific theories and experiments in the understanding of the physical nature of the world. Unfortunately nature is not an open book that can be easily read to enhance our knowledge, as it was always claimed by empiricists [7, 8]. What we already know dictates and is helpful in suggesting the experiments to perform. In other words, scientific hypotheses dictate the scientific experiments to be carried out, but experiments could also reveal something unprecedented, which needs to be interpreted, and this can be done by means of the imagination and intuition.

Could the observation of reality provide evidence for a theoretical hypothesis, which had been previously advanced? As mentioned above, the 20th century science appeared to be a knowledge mediated by a reasoning process that involves a conceptual construction. In fact, man should use models, hence, he has to propose some properties from which consequences are derived, and these predictions should agree with experimental data. An explanation must be attempted on the basis of a hypothesis, a conjecture on reality must be formulated, as stated by K.R. Popper (1902-1994) [9]. It is actually quite important to compare the conclusions of a theory with reality, because no physical theory can fully describe reality, which, as said above, remains richer than what man can build within his symbolic space. From this perspective we must appreciate how scientists seek new elements of truth showing their wish to grasp the rationality of reality [10]. Every scientist and A. Einstein certainly believed this [11], seeking for new elements of truth does not question the existence of an objective world, independent of our level of observation.

Well then, the previous question can be further clarified by the following: Can we say we have really described objects which we cannot actually see?

We would like to answer this question by proposing an example taken from Theobald's book about the philosophy of science [12]. We ought to observe that one of the most common philosophy objections to science's claim to be able to describe invisible objects is that it is not possible to assert with absolute certainty the existence of things which cannot be directly observed. Therefore, we have to look over the conditions that might lead scientists to make assertions about the reality of the objects they observe. The presumption of seeing something in our world should be verifiable: a) we see a horse; b) we see a molecule; c) we see an elementary particle. A different procedure could be implied, but the logic of seeing should be the same. Let us examine the logical structure of these two sentences: a) in the lake something moves, then there is a fish, we see a fish; b) in the August sky we see a trail of light, then a shooting star can be seen, we see a shooting star. These assertions are justified because in our background knowledge, a great many people have had experiences of going fishing as well as some elementary understanding of astronomy. If all these assertions are legitimate, then in the same way, detection and examination of traces in a bubble or cloud chamber allows us to state that there is an elementary particle, we see an elementary particle, because, in the scientific background, traces in a cloud chamber identify elementary particles (probably this statement is relatively comprehensible to "outsiders"). While watching the performance of a puppet show, we might not realize we are dealing with puppets. But if we know everything concerned with the art of the puppet theatre (this art being our cultural background), we admire the skills of the puppet operators, and then, watching puppets means seeing these operators performing an action.

Let us conclude by saying that without the strong conviction that our theoretical constructions achieve a representation of reality[3] (which is

---

[3] We are convinced that such a thing as truth exists and share the opinion of R. Bailey [13] that a truth-seeking approach assumes that our errors can be corrected and a progress made.

surely partial and improvable), there would not be scientific knowledge, only a mere practice. This enterprise to interpret "signs" coming from an unknown reality is a process which proceeds through questions about how things are made. The attempt to answer these questions, which is the scientific work, is also the way to seek truth (probably a partial truth). It is an endless process, because questions will never end, or rather they will end by generating other questions to be answered. Actually,

> *we are continuously present at the daily drama of the hard work of study, the cooperation and rivalry between theoretical advances and experimental facts and this endless clash of methods which is the dynamical character of our contemporary scientific culture* [16] (our translation).

REFERENCES

[1] Lecourt, D., & Canguilhem, G. (1969). *L'épistémologie historique de Gaston Bachelard*. Paris, France: Libraire Philosophique J. Vrin.

[2] Bachelard, G. (1971). *Epistémologie. Textes choisis.* (essays collected by Lecourt, D.) Paris, France: Presses Universitaires de France (PUF).

[3] Bachelard, G. (1949). *Le Rationalisme appliqué*. Paris, France: Presses Universitaires de France (PUF).

[4] Bachelard, G. (1932). Noumène et microphysique. *Recherche philosophique*, 1, 55-65.

[5] Roy, B. (2006). Noumenon and Microphysics. P*hilosophical Forum*, 37, 75-84.

Furthermore, we also think the term reality needs a better explanation, and we are aware of the quarrel in the present literature in which many positions are debated by historians and philosophers of science (naive realism, practical realism, referential realism, realism with a small 'r' or with a big 'R' and so on) [14]. We share the ideas of one the most famous Italian theoretical chemists of our time, G. Del Re (1932-2009). We met him once at the University of Studies of Catania and have attended and enjoyed his seminar about the connection between physical statements and reality. '*There are things, events and processes independent of our own existence and will, and they can be individually known by us, within the limits imposed by our senses and brain, as existing and distinct from other objects [...], therefore, within the frame of realism one should accept this conclusion: what we call the 'structure of a molecule (which we can define a second class entity) is in principle inherent to molecular reality, and hence real*' [15].

[6] Jaki, S. L. (1981). *Chance or Reality: Interaction in Nature versus Measurement in Physics.* Philosophia (Athens), 10-11, 85-102, and lately reissued as a book: Jaki, S. L. (1986). *Chance or Reality and Other Essay.* Lanham, MD: University Press of America & Intercollegiate Studies Institute.

[7] Enriques, F., & de Santillana, G. (1973). *Compendio di Storia del Pensiero Scientifico.* Bologna, Italy: Zanichelli.

[8] Mason, S. F. (1962). *A History of the Sciences.* New York, NY: Collier Books.

[9] Popper, K. R. (1963). *Conjectures and Refutations: The Growth of Scientific Knowledge.* London, UK: Routledge.

[10] Bachelard, G. (1967). *La formation de l'esprit scientifique.* Paris, France: Librairie Philosophique J. Vrin.

[11] Resnick, R. (1980). Misconception about Einstein: his works and his views. *Journal of Chemical Education,* 57, 854-862.

[12] Theobald, D. W. (1968). *An introduction to the Philosophy of Science.* London, UK: Methuen.

[13] Bailey, R. (2001). Overcoming Veriphobia - Learning To Love Truth Again. *British Journal of Educational Studies,* 49, 159-172.

[14] Vihalemm, R. (2011). The Autonomy of Chemistry: Old and New Problems. *Foundations of Chemistry,* 13, 97-107.

[15] Del Re, G. (1998). Ontological Status of Molecular Structure. *HYLE-International Journal for Philosophy of Chemistry,* 4, 81-103.

[16] Bachelard, G. (1972). *L'engagement Rationaliste.* Paris, France: Presses Universitaires de France (PUF).

Chapter 6

ABOUT THERMODYNAMICS

This sixth part is dedicated to a short introduction to the study of thermodynamics. We will avoid a mathematical approach, presenting instead this topic following the well-known book by I. Prigogine (I. V. Prigogine 1917–2003, Nobel Prize in Chemistry in 1977) and I. Stengers (1949-) [1]. We have often found that the teaching of thermodynamics[1], and the corresponding textbook chapters are restricted to the necessary details and do not give students the opportunity to understand the cultural break that thermodynamics represented with respect to Newtonian dynamics.

What was new and important about the introduction of thermodynamics?

The formulation of the laws of motion by Newton was a synthesis between the mathematical description of the motion of bodies (Keplero and Galileo): it is possible to describe the location, speed and acceleration of a

---

[1] Chemical thermodynamics has the central problem of facing chemical systems in equilibrium, while physical thermodynamics is generally interested in material behaviour. To achieve the task of explaining the behaviour of hundreds of thousands of chemical substances, physical chemists have used many models which have been refined and adapted to different situations by semi-empirical approaches coming from chemical concepts that are not reducible to physics [2]. Hence, chemical and physical thermodynamics are disciplines different enough from each other, because their tasks are different (though sharing the same principles, the same heat and work signs, the same variables as pressure, temperature and volume, and the same thermodynamic functions as internal energy, enthalpy, entropy and free energy). As asserted by A. Baracca:'*Chemists [..adopted..] a completely new thermodynamic approach, essentially replacing thermodynamics for mechanics as a basic reference frame to obtain concrete and general results, as well as to elaborate new conceptions*' [3].

body at any moment in time in a three-dimensional space by means of calculus. Newtonian physicists were mainly interested in the study of the acceleration of bodies as a function of the changes occurring in the system. This means the study of the different forces acting on all points of the system. The formula $F = m \times a$ which is the fundamental principle of dynamics, predicts equivalence between force and acceleration and represents the causal structure of dynamic systems: nothing happens if there is not the effect of a given force, in other words, there is no effect without a cause. Now, when the problem is represented by a one-body system, the study of its motion is very simple. On the contrary, when a multi-body system is considered, the problem is extremely complicated, because of the mutual influence of the motions of each body. Thus, the instantaneous state of the system at any given point in time is described by speed and acceleration of each body through a system of equations which needs, as starting points:

1. a set of forces among bodies, which are a function of their distance;
2. this means the body undergoes a particular acceleration when it finds itself under the influence of the attractive force of another;
3. the change in the relative distances among bodies is the effect which is obtained.

The integration of motion differential equations allows us to gain information on the trajectory of bodies in the system as a function of time. In other words, the achievement of a function $r(t)$, which describes the trajectory of the body under the influence of the others' forces. This is the key role of dynamics and every description involves two kinds of empirical data: a) the knowledge of the initial conditions, namely position and speed at the initial time, b) the knowledge of forces acting among bodies. In other words, in the study of the dynamics of a system, everything consists in the initial conditions of the problem. There is an intrinsic property of dynamics when describing a system: its equations are symmetrical by means of time reversal operation. Namely, a reversal of time (t in $-t$) or speed (v in $-v$) restores the conditions of the system to a situation identical to those in which it started. It is, so to speak, as if in a world governed by Newtonian dynamics, a movie may be shown backwards, to reignite an already extinguished match or to put back together the fragments of a broken china vase. However, we all know that this is a fiction. The world of dynamics considers these events to be reasonably possible in the same way as all the events we are used to watching in our world. But our experience contradicts this claim: surely we can all affirm that our world consists of irreversible processes. Living organisms, being subject to birth, growth, old age and death, seem to contradict this reversibility concept coming from the Newtonian dynamics and its generalization as well. During the second half of the 18th century, German philosophers of nature rejected this generalization because it was not in line with the complexity of the vital processes of life: everything in the universe cannot be simply explained in terms of mechanical laws [4, 5].

In order to answer to the question about the new ideas proposed by thermodynamics, it is possible to state that the claim of dynamics was challenged by the rise of thermodynamics:

heat became the rival of gravity [1].

In England there was a rapid diffusion of heat engines during the Industrial Revolution towards the end of the 18th century; in other words, the idea of exploiting the mechanical effect of heat developed. In fact, the initial drive to understand thermodynamics came from the need to increase the efficiency of steam engines for the conversion of heat energy into work (the nature of heat, however, was a question that arose many years later). Therefore, the initial motivation was to learn how to exploit heat energy for a technological use. The real problem to be tackled was the study of how heat can be converted into mechanical work, that is, which conditions should be satisfied to exploit heat energy and make steam engines work. Very often, it is common enough to be in a hurry to carry out industrial processing without having studied the pertinent scientific problem in depth, in this particular case, the knowledge about the nature and pathways of heat transfer. At the beginning of the 18th century, the tremendous success and power that England had achieved using steam engines, convinced S. Carnot (1796-1832),[2] a French engineer, to study the physical processes involved in the transformation of heat into work.

---

[2] As it was stated by M. Thalos [6], Carnot's assertions started from his idea that caloric, just as a flow of waters, flowed from hot to cold source.

Let us examine the salient points of this conceptual break with dynamics.

1. Heat transfer cannot be attributed to dynamic interactions among masses, it is only proportional to the temperature difference between two bodies. Everybody has a mass and is, therefore, subject to the force of gravity, but at the same time it is able to accumulate and transfer heat. It is the proportionality coefficient between the heat flux and the temperature difference that changes. And we all know that heat transfer law is irreversible, meaning that heat does not return spontaneously after it has been transferred [7].
2. It is always possible to act mechanically on a system by changing its pressure or volume, to exert a thermal action by supplying or removing heat, or to make a chemical action happen by burning something in the presence of air (which means letting a combustion reaction occur). Therefore, we are not observing trajectories or predicting them, but rather we would like to act directly on a system and observe how it reacts on changes coming from the outside.
3. The energy-saving principle, which in Newtonian dynamics is expressed by the constancy of the Hamiltonian (the sum of the potential and kinetic energies), has to be considered valid even in thermodynamics, but it needs to take a step forward. It was J. P. Joule (1818-1889) who, introducing the mechanical equivalent of heat, suggested the idea of an energy conversion: energy is always quantitatively saved while qualitatively changing its form. This last proposition is the first law of thermodynamics: heat can be converted into work and vice versa.
4. But

   It is also true no heat machine will give back to world the coal which has been consumed [1].

Even though energy is saved, at the same time there is a loss of useful energy because of its dissipation. For the first time scientists are constrained to introduce the concept of machine efficiency, because this dissipation causes a lack of energy balance. In order to face this problem the second law of thermodynamics was elaborated, accompanied by a new concept: entropy. It was first formulated by R. Clausius (1822-1888) a German physicist and mathematician, considered one of the founders of thermodynamics [4]. In Newtonian dynamics the ideal of reversibility coincides with energy saving. In the transformations of a physico-chemical system, energy is saved even if these transformations are irreversible. It is necessary to distinguish between useful energy transfers and those dissipated, which are inevitably lost when the machine is returned to the initial conditions. In fact, several of the transformations in a system are irreversible and the increase of entropy just indicates an irreversible evolution of the system. This means that the energy consumed was ultimately lost via waste heat, hence, limiting evolution of an isolated system towards a particular state. The entropy increase expresses that in the physical world there is 'an arrow of time' [8]. Therefore, entropy becomes the symbol of a nature which evolves in only one determinate direction and cannot come back, unless for a closed system there is significant amount of external intervention. For a dynamical system the memory of the initial conditions always persists, while for a thermodynamic system, once the equilibrium point has been reached and the maximum entropy obtained, there is no more memory of the initial conditions.

Two comments need to be made. First, there have been several attempts to extend the conclusion of the second law of thermodynamics, which as a scientific law has a fairly limited sphere of applicability (the behaviour of a physicochemical system in this case) to reality as such. This goes well beyond the proper sphere through philosophical considerations, about which there has been a great debate for many years. The authors of the second principle of thermodynamics attempted to apply it to the physical universe as a whole (considered as an isolated system), and came to the conclusion that the entropy of the universe should constantly increase going towards the state with the highest disorder, which means the thermal death and, thus, the end of the universe itself.

Most of general chemistry textbooks drew students' attention towards this last statement, with much emphasis, without worrying about the fact that it has no real applicability towards the universe. In fact, the thermal death requires that the universe has a past with low entropy content, but as of today, we do not have any reliable measurement of the entropy of the universe (neither do we know if a similar quantity exists and if it is possible to measure it in a complex system like the universe). Moreover, there is no way to know if the development of the universe occurred in a state of low or high entropy. We can only collect data on the visible portion of the universe, which is limited by the fact that the light speed is a constant and cannot assume any value. Therefore, the entropy of the universe is not an easily measurable quantity, or we could even expect it to be a fluctuating quantity with characteristics able to create the evolution of complexity [9].

Second, classical thermodynamics does not require hypotheses about molecular structure. Its results could be similarly obtained by considering a gas as composed by molecules or as being a fluid. In other words, classical thermodynamics obtains its results on the basis of the macroscopic properties of the system only, without committing itself to any particular microscopic model.

L. E. Boltzmann (1844–1906) introduced statistical methods for the study of physico-chemical systems, as he was convinced of the intimate atomistic nature of matter. Hence, these systems have to take a large amount of particles into account. He successfully searched for a molecular interpretation of the entropy increase within a thermodynamic system, considering the collisions occurring among the gas molecules. These collisions lead to a Maxwell distribution of the molecular speeds and the system behaves irreversibly [8, 10]. In other words, while each molecule (when considered isolated from the others) belongs to the Newtonian world, thus reflecting the physics of order, all the molecules considered as a whole (a large "crowd" of molecules) behave in a disorderly way. Therefore,

> *within a physical system it is the "crowd" that gets into trouble* [11]
> (our translation),

because all the molecules do not coherently align to point towards one useful direction, but rather they are scattered randomly in all directions.

Boltzmann's *H* function, which coincides with negative entropy, was derived on the basis of locations and speeds of particles and is always a decreasing or constant function of time. This caused a fierce contention in the community of physicists: how is it possible to derive information about irreversibility from Newtonian laws which by their nature dealt with symmetry (or reversibility) of time [12]? Moreover, how may time dependence arise from the non-time? J. H. Poincaré (1854–1912) a French mathematician and theoretical physicist, sharply criticized the work of Boltzmann and said that he would never recommended that his students read any of Boltzmann's papers, as their conclusions clash with the bases. And yet, Boltzmann is the founder of statistical thermodynamics because he tried to correlate the entropy increase in an isolated system with the increase of molecular disorder. Entropy of a system represents the amount of energy connected to the number of ways in which a particular partition may be achieved, meaning that the evolution of an isolated system is irreversible and takes it to the most probable state. Once equilibrium is reached, the macroscopic variables (pressure, volume and temperature) can fluctuate, but the system will try to remain in equilibrium. In other words, the equilibrium state acts like a spring which draws the system towards it, when the system is slightly further away from it. It is like a potential well in which the system will fall into sooner or later. This could be attained in closed systems (where the energy exchange is allowed) because entropy may decrease, and more ordered structures may be obtained, only changing the values of the thermodynamic variables. In this case the equilibrium state in a system results from a competition between enthalpic and entropic contributions ($G = H - TS$).

When considering open systems, which like cells, can exchange both energy and matter, it is not possible to think of isolating them from the environment, because living systems would die if they were isolated. In the world we know, equilibrium is a rare and precarious state. Therefore, direct extrapolation of thermodynamic laws, taken out of the context of their validity and scope, leads to the conclusion that biological life (where everything is directed towards a goal, despite the increase in entropy and molecular chaos) would be a rare event.

The very notion that matter tends to some absolute state of equilibrium runs counter nature itself; it is a lifeless and abstract view of the universe [8].

Thermodynamics gives an answer only for chemical reactions and physico-chemical transformations which are reversible. This answer is detailed in terms of energy dissipation, destruction of initial conditions and evolution of the system towards a more disordered state.

When a process is far from reaching equilibrium, there is a great chance that system instability occurs. Therefore, some fluctuations can be amplified up to the point in which the system itself no longer obeys thermodynamic laws and finds many different pathways to choose from. In other words, a strict correlation between distance from equilibrium and self-organization could be made. When dealing with open systems that are far from equilibrium conditions, energy and matter dissipation could help them to find a newly ordered state. The state now achieved is dependent on the previous history of the system. This same thing might happen to biological systems due to their complexity. Among the pertinent examples that could be proposed, there is the phenomenon of convection, the ordered movement of molecules within a fluid as a result of their being heated, or the connected redox chemical reactions which are named chemical clocks for their peculiar phenomena [13]. For microstructure situations, quite distant from equilibrium, it is possible to achieve a much greater organization as a result of a microscopical amplification. At the right moment, this could favour a specific pathway among all the other equally probable pathways.

But the question remains, what happens to intervene and cause that specific pathway to occur, at the right moment?

REFERENCES

[1] Prigogine, Y., & Stengers, I. (1979). *La Nouvelle Alliance. Métamorphose de la science.* Paris, France: Éditions Gallimard.

[2] van Brakel, J. (2000). *Philosophy of Chemistry.* Leuven, NL: Leuven University Press.

[3] Baracca, A. (1996). Chemistry's Leading Role in the Scientific revolution at the Turn of the Century. In V. Mosini (Ed.), *Philosophers in the Laboratory* (pp. 61-80). Rome, Italy: Musis, EUROMA.

[4] Mason, S. F. (1962). *A History of the Sciences.* New York, NY: Collier Books.

[5] Schummer, J. (2003). The notion of nature in chemistry. *Studies in History and Philosophy of Science,* 34, 705-736.

[6] Thalos, M. (2012). *The Lens of Chemistry.* Science & Education, 21, doi: 10.1007/s11191-012-9443-y.

[7] Bellone, E. (1978). *Le Leggi della Termodinamica da Boyle a Boltzmann.* Torino, Italy: Loescher.

[8] Prigogine, Y. (1991). The arrow of time. In: C. Rossi & E. Tiezzi (Eds.), Ecological Physical Chemistry. *Proceedings of an International Workshop* (pp. 1-24). Amsterdam, NL: Elsevier.

[9] Barrow, J. D. (1999). *Between Inner Space and Outer Space. Essays on Science,* Art, and Philosophy. Oxford, UK: Oxford University Press.

[10] Bonomo, R. P & Riggi, F. (1984). The evolution of the speed distribution for a two-dimensional ideal gas: A computer simulation. *American Journal of Physics,* 52, 54-55.

[11] Arecchi, F. T., & Arecchi, I. (1990). *I Simboli e la Realtà.* Milano, Italy: Jaca Book.

[12] Barrow, J. D and Tipler, F. J. (1986). *The Anthropic Cosmological Principle.* Oxford, UK: Oxford University Press.

[13] Lefelocz, J. F. (1972). The color blind traffic light. An undergraduate kinetic experiment using an oscillating reaction. *Journal of Chemical Education,* 49, 312-314.

Chapter 7

CHEMISTRY EDUCATION AND SCIENCE TEACHING

Our introductory lectures are generally proposed in form of free seminars during the semesters in which the General Chemistry program is completed, and they are opened for public comment to all science and chemistry students and academic staff of the Department of Chemical Sciences of our University. Everyone has the chance to put questions and make critical comments. We always get the impression that students attend these lectures with great interest, as witnessed by the frequent request for more information on these subject areas and, furthermore, for a receipt of a copy of the manuscript of each lecture, with the pertinent references. Not only, but the students' motivation and participation to the regular didactic activities of the General Chemistry course have been greatly stimulated. Frequently, during the semesters some of the assertions made in the introductory lectures (or made at the end of that particular topic) are reintroduced in order to show the connection with the topics under study and on many occasions students are helped as much as possible to express their doubts directly or present questions. Some years ago one of us was most impressed by a student[1] (also working as technician in the physico-chemical laboratories of our department) who, after having attended the lecture about thermodynamics, some days after, called him to watch at the Belousov's oscillating reaction [1] he had made in his lab; the repetitive

---

[1] This student, whose name was Marcello Caruso (1968-1992), died a few months later in consequence of a heart attack caused by a badly treated case of pleurisy.

changes of the red and blue colours of ferroin iron complex could be seen in times which varied as a function of the stirring rate of the solution.

Now let us briefly consider about the way scientific topics are usually covered and presented to our students nowadays. We would like to start by stressing in an emphatic statement by G. Bachelard [2], who affirmed that the goal of scientific teaching is often reduced to the description of facts. This idea is reminiscent of 'memory empiricism' and prevents students from seeking rationalizations and raising questions [3]. On the contrary, it is very important to comprehend 'the order of reasons', because it trains students to think rationally. An analogous assertion was made by M. Niaz:

> *[A] lack of conceptualization clearly shows as [students] simply memorize the scientific laws and the corresponding mathematical equation and [only] look for values to plug in* [4].

He called this habit 'algorithmic mode' of teaching and learning of scientific disciplines.

The majority of General Chemistry textbooks present scientific knowledge and fundamental notions dealing with the subject as if all that we know is definitively achieved. There is no space for criticism or rethinking of the results that are presented. Many science lecturers at the most important universities only worry about the teaching technical details of their scientific discipline. We could use some of the assertions of T. Kuhn (1922–1996) [5],[2] to affirm that it is as if students were always put in the presence of the modern scientific 'paradigm',[3] but they are not aware of the problematic nature of the 'pre-paradigmatic' period.

> *tend to make scientific revolutions invisible* [7],

---

[2] The figure of T. Kuhn is surely most controversial, because his views of the history of science have had a great impact on science, philosophy of science as well as on science education. However, his statements are not universally appreciated and a great debate has developed on his accounts of the nature of science and the consequent historical and philosophical considerations. To this purpose it could be interesting to read articles about Kuhn's views and their consequences on scientific education by H. Andersen [6], P. Slezak [7], B. Van Berkel et al.[8] and M. R. Matthews [9].

[3] The word paradigm needs a specification, because it is an ambiguous term as shown by R. Bailey [10] or P. J. Wendel [11]. Therefore we have to specify that we use it as a synonymous term for a working programme or a conceptual scheme which helps scientists in their work.

Still, textbooks considering only a developmental structure which shows the 'paradigm' components as aims, theories, concepts, methods, techniques and criteria. Moreover, lecturers and teachers have to consider that the power of history is to inform the present and that

> *history [of science] counteracts the scientism and dogmatism that are commonly found in science texts and* classes [12].

This is why General Chemistry textbooks in every country are so similar. Our students do not learn problems, questions or debates which were important in the past and could be relevant today (as we think), but the solutions that are now universally accepted.

> *Instead of learning answers to possible questions, students must learn to put questions to which they might find possible answers [...], they must be able to recognize problems that can be tackled with chemical knowledge* [13].

When learning a scientific curriculum, students meet neither the research front [nor] the history or foundation of a discipline, and, as we said above, textbooks give generally a distorted picture of the history of a discipline. Moreover, standard chemistry education is characterized by the phenomenon of educational positivism. Science is presented as an activity bringing about a constant progress in knowledge; not only, but also the scientific method is presented as the only way to investigate physical or human reality and, therefore, ignoring that scientists use a plurality of methods and approaches.

In all the scientific textbooks

> *positivism still appears to reign uncontested* [8].

We believe that teaching of scientific topics can be greatly aided by directing the student's attention to how scientific notions and practices have developed historically. Teaching of scientific topics cannot be reduced to merely necessary technical details, but they should be an extensive work directing the students' attention to an understanding of the cultural aspects of the main points of the course. It does not mean to teach a parallel course of the History of Science (in particular, History of Chemistry; it should be done, too! And if done, it must be coordinated with the other chemical disciplines). We are aware of the difficulty in convincing our colleagues that a change would be desirable and that the main aim of teaching chemistry is not to cover the most common topics of the discipline during a course. We think it is better to reasonably limit the number of topics to explore (nowadays the hours in a semester are never sufficient) and illustrate only some of them, but considering them historically and philosophically. A cultural approach, which is appropriate at a university level, needs some historical and epistemological arguments about how to achieve the insights necessary for a deep understanding of the topics discussed. As pointed out by G. B. Kauffman:

> *integrating the history of science in our courses [...] can be done extensively throughout the entire course or only for those topics in which the instructor has a particular interest.[...] The education of a chemist without some inclusion of history remains unsatisfactory and incomplete* [14].

Introductory science courses should be designed with the explicit goal of introducing students to the main points of the history and philosophy of science, because

> *all introductory science courses unavoidably serve as vehicles of general education. Course designers ought to take this obligation seriously* [15].

Today a research scientist, rather than being an innovator, is a solver of puzzles in the Kuhnian point of view (normal science): he concentrates his/her efforts upon solving puzzles, because the solution of these puzzles is the way indicated by the existing scientific tradition (paradigm). This fact inevitably has its counterpart in science education, because students are normally educated to become solver of puzzles and to accept the currently scientific results in an uncritical fashion. However, teaching students to solve problems and puzzles is not necessarily a bad thing, given their ordinary scarce attitude to apply themselves to their studies (just in the same way as finger exercises are used in training musicians, but these exercises do not prevent them to become fine musical interpreters or original composers). This teaching should always deal with the aim for understanding and comprehension of the subject; namely, what students *need to do is to learn concepts* [16] and are *required to acquire conceptual understanding* [4], rationality and ability to give of reasons.

Learning science involves today a process of enculturation into the ideas, models and methods of our conventional science. Actually,

> *the growth of scientific understanding goes hand in hand with initiation into a scientific tradition* [12].

As stated by K.R. Popper:

> *all teaching on the university level [...] should be training and encouragement in critical thinking* [17].

Therefore, it is proper to meet this requirement that it is important to break the routine involved in normal science education, by giving the students the opportunity to reflect upon the key points of a scientific discipline. These outlines of introductory lectures here presented, could be helpful examples for many teachers and lecturers who wish to make their teaching more effective. Certainly, in order to do this, many lecturers will be constrained to inquire about the historical and philosophical aspects of their discipline, and this work could be incredibly interesting. In other words, our students have to learn to appreciate the history of science and its connections with values and culture of society, its philosophical and metaphysical implications, in order to have the possibility to enrich their culture and human lives. This way students will be taught how to properly use their reasoning skills.

On the other hand, how could we teach without our personal and continuous study and research activity? Every great teacher has his own teaching style that is more than just knowing the subject matter, and would make clear the path he wishes his students to follow. He would involve himself in this path and be willing to accompany his students during their study in detail. For this reason, in proposing this path, it is as if he would go through it again, always starting from how he had understood the specific concepts encountered in the study of the discipline topics, studying them in depth and clarifying all the points which are not yet clear. Actually, he is proposing a method of approaching a particular reality (in our case Chemistry), even if this method is specific and cannot be extended to grasp the whole reality: every object requires a particular method to be used in order to allow for its knowledge [18]. In other words, it is an effort to raise student's awareness about the nature of science and educate them to recognize signs (for example, traces of experimental apparatuses), because it is through the interpretation of these signs, by means of the logical and mathematical apparatus which has been developed, that a knowledge of reality can be acquired by man. However, it is possible to know things only if motivated by an interest in them. Therefore, the student's goal is to identify himself with the reasons proposed by his teacher and take possession of this knowledge in a deep mood.

Furthermore, we will add a final consideration starting from the following questions: would the teaching of a scientific discipline only deal with optimizing historical, philosophical and educational elements with its specific technical details? Or perhaps, might the human factor elements also affect our way of teaching a scientific discipline? We would like to argue that the human factor is often central to any serious attempt to renew the didactical structure of our science curricula. After all, every education activity always implies some kind of a positive human relationship between teachers and students. We were fascinated by Einstein's pleasant recollection [19] of his friend, P. Ehrenfest (1880-1933), because it implicitly answers this question: what sort of person is a university or high school teacher?

> *Our revered master, Lorentz [...] had recognized Ehrenfest for the inspired teacher that he was and recommended him as his successor. A marvellous sphere of activity opened up to the still youthful man. He was not merely the best teacher in our profession whom I have ever known; he was also passionately preoccupied with the development and destiny of men, especially his students. To understand others, to gain their friendship and trust, to aid anyone embroiled in outer or inner struggles, to encourage youthful talent- all this was his real element, almost more than immersion in scientific problems. His student and colleagues in Leyden loved and esteemed him. They knew his utter devotion, his nature so wholly attuned to service and help.*

REFERENCES

[1] Lefelocz, J. F. (1972). The color blind traffic light. An undergraduate kinetic experiment using an oscillating reaction. *Journal of Chemical Education,* 49, 312-314.

[2] Bachelard, G. (1971). *Epistémologie. Textes choisis.* (essays collected by Lecourt, D.) Paris, France: Presses Universitaires de France (PUF).
[3] Souque, J.-P. (1988). The Historical Epistemology of Gaston Bachelard and its Relevance to Science Education. *Thinking: The Journal of Philosophy for Children,* 6, 8-13.
[4] Niaz, M. (2008). *Teaching General Chemistry: A History and Philosophy of Science Approach.* New York, NY: Nova Science Publishers.
[5] Kuhn, T. S. (1962). *The Structure of Scientific Revolutions.* Chicago, IL: University of Chicago Press.
[6] Andersen, H. (2000). Learning by Ostension: Thomas Kuhn on Science Education. Science & Education, 9, 91-106.
[7] Slezak, P. (1999). Does Science Teaching Need History and Philosophy of Science?. In G. Nagarjuna (Ed.), *International Workshop on History of Science: Implications for Science Education* (pp. 21-38). Mumbai, India: HBCSE.
[8] Van Berkel, B.; De Vos W.; Verdonik, A.H. & Pilot, A. (2000). *Normal Science Education and its Danger: The case of School Chemistry. Science & Education,* 9, 123-159.
[9] Matthews, M. R. (2003). Thomas Kuhn's Impact on Science Education: What Lessons Can Be Learned?. *Science & Education,* 88, 90-118.
[10] Bailey, R. (2006). Science, Normal Science and Science Education - Thomas Kuhn and Education. *Learning for Democracy,* 2, 7-18.
[11] Wendel, P.J. (2008). Models and Paradigms in Kuhn and Halloun. *Science & Education,* 17, 131-141.
[12] Matthews, M. R. (1994). *Science Teaching. The Role of History and Philosophy of Science.* New York, NY: Routledge.
[13] Schummer, J. (1999). Coping with the growth of Chemical Knowledge. Educación Quimica, 10, 92-101.
[14] Kauffman, G. B. (1991). History in the Chemistry Curriculum. In M. R. Matthews (Ed.), *History, Philosophy and Science Teaching, Selected Readings* (pp. 185-200). Toronto and New York: OISE Press.

[15] Earley, J. E. (2004). Would Introductory Chemistry Courses Work Better With a New Philosophical Basis?. *Foundations of Chemistry,* 6, 137-160.
[16] Nersessian, N. J. (1991). Conceptual Changes in Science and in Science Education. In M. R. Matthews (Ed.), *History, Philosophy and Science Teaching: Selected* Readings (pp. 133-148). Toronto and New York: OISE Press.
[17] Popper, K. R. (1970). Normal Science and its Danger. In I. Lakatos & A. Musgrave (Eds.), *Criticism and the Growth of Knowledge* (pp. 51-58). London, UK: Cambridge University Press.
[18] Giussani, L., & Zucchi J. (1997*). The Religious Sense.* Montreal: McGill-Queen's University Press.
[19] Einstein, A. (1950). *Out of my later years.* London, UK: Thames and Hudson.

Chapter 8

Conclusion

We presented a set of historical and philosophical considerations conceived to help students improve their understanding of the complex nature of science and, in particular, we illustrated how these considerations could be applied to some of the principal topics of a General Chemistry Syllabus of a first-level university course.

We started from the examination of the relationship between language of science and ordinary language and noted as science cannot be done using every day ordinary language, because new entities have to be defined in order to establish cause-and-effect relationships among disparate situations. We stated as there is a need for a mathematical approach to scientific problems, which consists in studying of models with strong emphasis on their controllability in a comparison with reality, and showed how this approach enriches our understanding of the reality of physical world.

Then we tried to answer the questions: what is chemistry about? What are the objects studied by chemistry?

We asserted that a chemist has put a lot of effort into recognizing the hidden order of chemical features in the natural world, or, in other words, has been in desperate search of a rationale concerning the phenomena which are present in the mineral kingdom. Unfortunately, these relationships are not immediately comprehensible and, at first sight, nature gives the impression of a great disorder. Therefore, the first step in applying the rational method was the abandonment of the natural kingdom as a place in which chemical processes can be explored and the use of the chemical laboratory to understand simple chemical reactions. Analysis, separation and purification were important steps which led to modern Chemistry, as well as chemical formulas like an attempt to show matter within geometrical schemes and energy exchanges as the way to understand of matter transformations.

After, an outline of a brief history of the valence concept was presented, beginning from the Laurent's theory of types through the Frankland's valence theory up to the modern concept of the chemical bond. Some considerations were proposed on the meaning of the invariable Kekule's four-valence of carbon, from which Organic Chemistry developed. Ionic and covalent bonds were seen as steps along this road, even if the concept itself of the chemical bond itself has been questioned as unnecessary by theoretical chemists interested in exploiting molecular properties. In this perspective, important considerations on the connection between the notion of reality and its intelligibility as proposed by A. Einstein were laid out.

Neither did we disdain to analyse the historical development of the rational organization of simple substances in nature. The merits of Mendeleev's organization of the periodic table by cross-ordering atomic mass and valence were largely discussed and a comparison of his periodic table with our modern periodic table was also made. Moreover, some ideas were presented on the atomic number notion as conceived after the Moseley's pioneering work and a way to distinguish among physical, chemical and nuclear transformations was drawn out.

Introductory remarks on the atomic theory were discussed to give evidence for all the main points that are crucial from a cultural point of view, without entering into the mathematical details of quantum mechanics. We also explained how science in modern times is a collective activity rationally organized, and considered the connection between achievements in theory and reality of the objects under study, putting the right emphasis on the conclusions reached by science. We also discussed considerations about the cultural consequences of Heisenberg's uncertainty principle and the way scientists are able to make assertions about reality, bringing into focus the particular world they experience.

Stimulating considerations have been made on the way of teaching thermodynamics and criticism was advanced against the corresponding chapters of university General Chemistry textbooks that are often restricted to the necessary details only. As it was shown, it is almost important to give the students the opportunity to comprehend the cultural break with the Newtonian dynamics that thermodynamics represented.

Some remarks about the contemporary teaching of scientific disciplines were also given. Our point of view on how to improve the teaching of Chemistry and in general of science at the high-school or first-level university teaching courses was proposed, not only on the basis of our teaching experience, but also exploiting the conclusions reached by many scholars involved in educational studies.

But let us make some final comments regarding some of the issues that were raised.

As we stated at the end of chapter 7, the ability to know anything that we choose to know follows different methods and paths because our way of reasoning is agile and polyvalent. In fact, there is not usually one single method of knowledge, but a plurality of methods and approaches, so that choice is inherently dictated by the characteristics of the particular thing we are going to know. In other words, the method of gaining knowledge about the world is determined by the object under investigation. There some aspects of our life like friendship, loyalty, fairness and love that cannot be known by using a mathematical or physico-chemical approach, namely strictly applying the rigorous scientific method (given that there was only one [1]). And nonetheless we know our world and our whole aim is to reach certainty.

It is important to think back over the concept of reason we have attained from our cultural tradition. Therefore, what we really need is to overcome this modern concept of reason, which looks at reasoning as a way to set limits to reason or thought: reason as a cage which if anything doesn't come in, this thing doesn't exist. We are called to "broaden" our reason in order to embrace possibly even those events which could turn out to be obscure, because this is the only way to improve our knowledge.

Reality of our world (about things that we observe or friends whom we live with) always presents itself through signs, therefore, the interpretation of signs is extremely important in doing science, in other words it is a way of approaching the physical objects. Just considering all peaks, curves, trajectories that come out from experimental apparatuses and would hopefully allow us to say something about a reality which we are not able to see. This experience of the interpretation of signs is also valid for humans, in fact human existence is already a sign and points to something beyond itself [2]. And it emphasizes that there must be an answer beyond the material aspects of existence: why is there something rather than nothing? Reality is there and, after all, was not created by us.

Doing science is an intellectual work by which man tries to understand how things are made, how they behave and build up pictures, representations of reality, which are closest approximations to the truth. As cited above, A. Einstein stated that physical concepts are free creations of human mind and are not, however it may seem, solely determined by external world and we believe that they help us to reconstruct a true representation of reality [3].

Galileo (not very many people know that he was a good Christian) said that God wrote the book of nature in the form of the language of mathematics. Now through the language of mathematics we are able to infer the universe structure and truly understanding nature, and mathematics is an invention of the human being.

> *It seems to me almost incredible that an invention of the human mind and the structure of the universe coincide. Mathematics, which we invented, really gives us access to the nature of the universe and makes it possible for us to use it. Therefore, the intellectual structure of the human subject and the objective structure of reality coincide: the subjective reason and the objective reason of nature are identical.*

It was the answer of Pope Benedetto XVI to a high school student who asked during a question-and-answer session on April 2006 why faith and science are generally presented as enemies against his wish to see them within a deep unity [4]. And these assertions are not really different from those by A. Einstein reported at the end of chapter 5.

It is important to recall that modern science got its start by operating a reduction of reality, carving out "slices" of reality that were possible to examine and explain, and leaving out all the rest intentionally. This reductionism has as consequence that all the assertions of science, of a particular science, get their validity within the borders of the reality under study, so as every pretension of scientists to tell people something about the whole reality should be considered inappropriate and therefore at least awkward. As affirmed by P. K. Feyerabend:

> *[...] scientific knowledge is too specialized and connected with too narrow a vision of the world to be taken over by society without further considerations. It must be examined, judged from a wider point of view that includes human concerns and the values flowing therefrom, and its claims to reality must be modified so that they agree with these values* [5].

We would like to affirm that scientism is not science, but only a philosophical and ideological point of view which has little to do with scientific knowledge, because it reduces human reason in the habit of considering true only those experiences that can be expressed within the language of mathematics. And this could go well (and it would have been really the simplest thing) if we could observe our universe to be homogeneous. Unfortunately, of course the situation is different and the universe where we live is not homogeneous and this inhomogeneity can be resolved by means of neither postulates nor a firm faith (and scientism has certainly the characteristics of fideism):

> *facts cannot be ignored and have to be accepted even when they cannot be understood.* [6] (our translation).

The knowledge processes always imply the occurrence of an event that would have not been expected and cannot be connected to antecedent events, something new and unpredictable that appears in our life or in our study and research, and is capable to break our intellectual schemes associated with what has been already known. As it often happened in the history of science, the occurrence of new observations let scientists confirm or refute the physical picture of what has been already known. And as, after all, it is happening these days in the scientific community after the experimental discovery of the Higgs boson in the Large Hadron Collider (LHC) at Geneva, where the experiment on the collision among protons at the highest energies, after about sixty years, seems to confirm the Standard Model, one of the most relevant theories of physics.

G. Bachelard thought and affirmed with great vigour:

> *"We can have no a priori trust in the knowledge that the immediately given pretends to give us. It is neither a judge nor even a witness; it is an accused whom, sooner or later, we will convince that he lied. Scientific knowledge is always the reform of an illusion"* [7, 8].

The Jewish French philosopher A. Finkielkraut [9] stated many times that if we are not able to save something that happened, the consequence will be like losing contact with reality. Therefore, when refusing reality as it appears to our sight, man does not try any more to form his reason accepting the image of the world and creates a fantasy world to replace reality or forces the world to fit better within his reasoning schemes. The occurrence of an unpredictable event, once recognized as such, happens with a surprise and breaks down our prejudices. Accepting a new event and the surprise that comes with it, allows us to overcome the well-known knowledge, to succeed in broadening our concept of reason and to better understand the natural world. Actually, reason comes out from its cage and lastly gets down to work.

Anyway, since the research never ends and there are always questions to answer, reason achieves its truly real purpose not only as long as does not ignore facts and lets itself be guided by experience rather than prejudice, but also when it allows man to get answers to the relevant questions like "from where?" and "towards where?".

REFERENCES

[1] Feyerabend, P. K. (1988). *Against Method. Outline of an Anarchistic Theory of Knowledge.* London, UK: Verso.

[2] Giussani, L., & Zucchi J. (1997). *The Religious Sense.* Montreal, Canada: McGill-Queen's University Press.

[3] Einstein, A. (1950). *Out of my later years.* London, UK: Thames and Hudson.

[4] Magister, S. *Faith By Numbers. When Ratzinger Puts on Galileo's Robes.* (2008) Available from: http://chiesa.espresso.repubblica.it/articolo/213210?eng=y

[5] Feyerabend, P. K. Galileo and the Tyranny of Truth. (1985) *The Galileo affair: A meeting of faith and science. Proceedings of the Cracow Conference,* May 24-27, 1984, Città del Vaticano: Specola Vaticana (edited by Coyne G.V.), p 155-166.

[6] Borghi, C. (1976). *Se volessimo vederci chiaro*, Milano, Italy: Jaca Book.

[7] Bachelard, G. (1971). *Epistémologie*. Textes choisis. (essays collected by Lecourt, D.) Paris, France: Presses Universitaires de France (PUF).
[8] Roy, B. (2006). *Noumenon and Microphysics*. Philosophical Forum, 37, 75-84.
[9] Finkielkraut A. *Interview given to S.M. Paci, 30GIORNI, International magazine edited by G. Andreotti, Rome.* (1998) Available from: http://www.30giorni.it/articoli_id_15525_l1.htm

ACKNOWLEDGMENTS

The authors are greatly indebted to their students who have had the patience to hear out these considerations about some historical and philosophical problems of General Chemistry topics and remarks on the scientific work. They have showed a great enthusiasm for availing themselves of the opportunity to be aware about the cultural themes that scientific problems normally introduce in our today's life.

The authors also thank Dr. A. Tontini, professor of Pharmaceutical Chemistry at the Department of Biomolecular Sciences, University of Studies of Urbino, Italy, Dr. A. Lanza, senior researcher, INAF Catania, Italy, Dr. G. Villani, senior researcher, ICCOM Pisa, Italy, for the patience with which they had red this manuscript and for the pertinent suggestions had kindly made to improve it, and Dr. V. Mosini, professor of Chemistry, Society and Sustainable Development at the Department of Chemistry, University of Rome "La Sapienza", Italy, for the kind gift of a copy of the book "Philosophers in the Laboratory", collecting the proceedings of the meeting held in Rome, Italy, 1-2 December 1994.

R. P. Bonomo is greatly indebted to PhD Monsignor F. Ventorino, professor of Philosophy at the St. Paul Theological Institute of Catania, Italy, for his friendship, fatherly care, relevant advices and teaching of Philosophy, in particular, Philosophy of Science. Many years ago, his question posed to the young scientist: "Do you know what this thing that we called science really is?" was revolutionary. Thanks to his efforts R. P. Bonomo, together with other colleagues involved in the study of scientific disciplines, was introduced to the reading first of the Popper's philosophical position, in particular, of the refutability criterion to be applied to scientific theories, second the historical deliberations of Kuhn with the concepts of "revolutionary science", "normal science" and "paradigm", and third the Bachelard's philosophy of science, in particular, the "epistemological break" and "epistemological obstacles" categories, which underline a real discontinuity in the history of science. It was a real hard work, but very satisfactory and profitable in the long term, because it allowed him to deepen the discussion about some cultural aspects of the Chemistry development.

R. P. Bonomo would also like to thank Dr. F. Riggi, professor of Experimental Physics and nuclear physics scientist at the Department of Physics and Astronomy, University of Studies of Catania, Italy, for his patient friendship and all those continuous and passionate discussions about reality of the physical world and Dr. E. Ciliberto, professor of General and Inorganic Chemistry and estimated researcher within the field of studies relating to art and archaeology, at the Department of Chemical Sciences, University of Studies of Catania, Italy, with whom he shares his love for teaching Chemistry.

The authors would like to extend their thanks to all the referees of Science & Education journal for what they have done for them. The content of this book (in a shorter version, of course, as an article) was previously submitted for publication to that journal and three of the four reviewers appreciated it and helped them to improve their manuscript in a way that is quite uncommon in the scientific community. The authors really say there were no words to express their gratitude. Fortunately at the end of the blind peer review process, the editors did not feel to take the responsibility of allowing the paper for publication and this decision let the authors think of the better chance of having a booklet published. For the same reason, the authors wish to thank Dr. Liberato Cardellini, professor of Chemical Fundaments of Technologies at the Department of Science and Engineering of Matter, Environment and Town Planning, Polytechnic University of Marche, Italy. He was consulted as member of the Advisory Board of Chemistry Education Research and Practice and Journal of Science Education, and part of the Editorial Board of the Eurasian Journal of Physics and Chemistry Education. After careful consideration and in-depth analysis of the complexity of these notes, he had really encouraged the authors to publish them.

The authors would also like to thank CIRCSMB (Consorzio Interuniversitario di Ricerca in Chimica dei Metalli nei Sistemi Biologici) for supporting this work.

Thanks are also due to Miss Emily Antenucci for her editing of the final version of the draft of this booklet.

INDEX

A

algorithmic mode, 56
Analysis, 13, 63
Andersen, 56
Antenucci, 73
Arrhenius, 22
arrow of time, 49
atomic number, 32, 33, 34, 35, 64

B

Bachelard, 11, 16, 28, 34, 38, 56, 67, 72
Bader, 24
Bailey, 42, 56
Baracca, 45
Béguyer de Chancourtois, 28
Belousov's oscillating reaction, 55
boiling point, 21
Boltzmann, 50
Bonomo, iii, 71, 72
Boyle, 7, 8

C

Cardellini, 72
cardinal number, 32
Carnot, 47

Cavendish, 13
chemical bond, vii, 19, 21, 22, 23, 24, 64
chemical experiences, 35
chemical phenomena, 16
chemical reactivity, 21, 22
Chemical-Bond Theories, 22
chemistry education, 57
CIRCSMB, 73
Clausius, 49
cold fusion, 6
concept of reason, 65, 68
Couper, 14

D

Del Re, 43
density, 21, 24, 27, 35
dynamics, 46, 47, 48

E

educational positivism, 57
educational studies, viii, 65
Ehrenfest, 60
Einstein, 25, 42, 60, 64, 66
eka-alluminium, 31
eka-boron, 31
eka-manganese, 31
eka-silicon, 31

Electrochemistry, 22
energy exchanges, 16, 17, 64
entropy, 45, 49, 50, 51
epistemological break, 72
epistemological obstacles, 72
equilibrium, 45, 49, 51, 52

F

Feyerabend, 21
Finkielkraut, 68
formula, 13, 14, 15, 21, 23, 46
Frankland, 20, 21, 23, 64

G

Galileo, 45, 66
Gas Kinetic Theory, 22
geometry, 15, 38, 39
Gillespie, 24
gold, 34, 35
gravity, 47, 48

H

Hargittai, 24
heat, 12, 28, 45, 47, 48, 49
Heisenberg, 23, 37, 39, 40, 64
History of Science, 57
human factor, 60
Hund's rule, 24

I

inorganic compounds, 19, 21

J

Jaki, 40
Joule, 48

K

Kauffman, 58
Kekulé, 14, 15, 21
Keplero, 45
knowledge of reality, 1, 60
Kuhn, 56, 72
Kutzelnigg, 23, 24, 25

L

language of mathematics, 5, 66, 67
Lanza, 71
Laurent, 19, 64
Lavoisier, 11, 12, 13
Le Bel, 22
lead, 34, 35, 42
Lecourt, 38
Lewis, 23
Living organisms, 47
living systems, 51
Lorentz, 60

M

Mariotte, 7
mass number, 34
material substance, 38
materialism, 16
mathematical approach, 45, 63
mathematical model, 4
Matthews, 56
Maxwell, 50
mechanical equivalent, 48
mechanical work, 47
melting point, 21
memory empiricism, 56
Mendeleev, vii, 27, 28, 29, 30, 31, 32, 33, 64
method of knowledge, 65
Meyer, 29
microscopical amplification, 52
Molecular Spectroscopy, 16, 22
Moseley, 33, 64

Mosini, 71
multi-body system, 46

N

Newlands, 28
Newtonian dynamics, vii, 45, 46, 48, 49, 65
Niaz, 56
normal science, 72
nuclear experiences, 35

O

order of reasons, 56
ordinary language, 1, 2, 3, 4, 5, 63
Organic Chemistry, 14, 20, 21, 64
organic compounds, 14, 21, 22

P

paradigm, 56, 57, 58, 72
pathway, 52
Pauli's principle, 24
Periodic Table, 27, 31
Perrin, 29
philosophy of science, 11, 42, 56, 58, 72
physical experiences, 35
physical world, vii, 7, 32, 49, 63, 72
Poincaré, 51
Popper, 41, 59, 71
Prigogine, 45
proto-language, 1
purification, 13, 64
purity degree, 12

R

rational organization, vii, 11, 16, 64
realism, 43
refutability criterion, 71
Riggi, 72
Roy, 38

S

Scerri, 24, 27, 31, 32, 33
Schrödinger, 24
Schummer, 11
science, vii, viii, 5, 6, 7, 12, 14, 16, 19, 21, 37, 38, 40, 41, 42, 43, 55, 56, 57, 58, 59, 60, 63, 64, 65, 66, 67, 71
science education, 56, 58, 59
scientific discipline, 56, 59, 60
scientific knowledge, 7, 43, 56, 67
scientific law, 49
scientific method, 57, 65
scientific revolution, 37
scientific theory, 5, 37
separation, 13, 64
Slezak, 56
Stengers, 45
structural formulas, 14, 15
symbol, 3, 5, 49

T

Thalos, 47
Theobald, 42
Thermodynamics, 16, 22, 45, 47, 48, 49, 51
Tontini, 71

V

valence, vii, 19, 20, 21, 22, 23, 24, 30, 31, 33, 64
Valence Theory, 20
Van Berkel, 56
Van't Hoff, 22
Ventorino, 71
Villani, 71

W

Wendel, 56